Particle Tracking Velocimetry

Particle Tracking Velocimetry

Dana Dabiri and Charles Pecora
Department of Aeronautics and Astronautics, University of Washington, Seattle, USA

IOP Publishing, Bristol, UK

ISBN 978-0-7503-2203-4 (ebook)
ISBN 978-0-7503-2201-0 (print)
ISBN 978-0-7503-2202-7 (mobi)

DOI 10.1088/978-0-7503-2203-4

Version: 20191101

IOP ebooks

British Library Cataloguing-in-Publication Data: A catalogue record for this book is available from the British Library.

Published by IOP Publishing, wholly owned by The Institute of Physics, London

IOP Publishing, Temple Circus, Temple Way, Bristol, BS1 6HG, UK

US Office: IOP Publishing, Inc., 190 North Independence Mall West, Suite 601, Philadelphia, PA 19106, USA

To my family who allowed me the time and space to bring this book to fruition, and to my advisor, Morteza Gharib, whose guidance throughout my graduate years still guides me to this day.

–Dana Dabiri

For my grandfather, Dick

–Charle's Pecora

Contents

Author biographies

Dana Dabiri

Dana Dabiri is Associate Professor at the William E Boeing Department of Aeronautics and Astronautics at the University of Washington, in Seattle, where he also holds an adjunct appointment at the Mechanical Engineering department. He received his BS in Mechanical Engineering at the University of California, San Diego in 1985, his MS in Mechanical Engineering at the University of California, Berkeley in 1987, and his PhD in Aerospace Engineering at the University of California, San Diego in 1992. He was a post-doctoral researcher at UC San Diego from 1992–1993, thereafter becoming a post-doctoral research and research scientist at Caltech and until the end of 2001. In 2002, he joined the faculty at the William E Boeing Department of Aeronautics and Astronautics as an Assistant Professor, and was promoted to Associate Professor with tenure in 2009.

Dr Dabiri's work pursues innovating and implementing new and novel experimental methods to quantitatively visualize fluid flow in order to reveal new in-depth fundamental understanding of fluid mechanics that otherwise could not be determined. This interest was initially developed and honed during his graduate work at UC San Diego, where he developed the digital particle image thermometry and velocimetry (DPITV) technique, a novel experimental method that allows for simultaneous measurements of temperature and velocity within a plane, using micro-encapsulated liquid-crystal particles as the velocity and temperature sensors. During his post-doctoral research at Caltech, Dr Dabiri developed a method whereby free-surface deformation and near-surface velocities could be measured simultaneously, and used it to study the interaction of a vertical shear layer with a free surface. As a research scientist at Caltech, Dr Dabiri was part of a team that founded the defocusing digital particle image velocimetry technique, a method that allows for true 3D volumetric velocity measurements. Upon coming to the University of Washington, Dr Dabiri innovated on a novel approach for implementing a microscopic 3D particle tracking velocimetry (PTV) measurement system, as well as 2D and 3D PTV algorithms. At present, Professor Dr Dabiri's research activities are focused on further innovating 2D and 3D digital PTV algorithms, global wall shear stress measurements, and global non-intrusive pressure and/or temperature measurements within flows with the goal of using them, either singly or in combination, towards applications in bio-flows, aero-acoustics, heat transfer, combustion, and computational fluid dynamics modeling.

Charles Pecora

Charles Pecora studied mechanical engineering at the University of Miami, where he was introduced to fluid dynamics during a project focused on testing and optimizing a novel vertical axis wind turbine design. After graduating in 2016, he then pursued a Master's degree in Aeronautics and Astronautics at the University of Washington under the mentorship of Professor Dabiri, where he developed post-processing algorithms for particle tracking velocimetry. Since his graduation in 2019, he has become a systems engineer for APiJET, a small company that integrates and analyzes real-time aircraft data to improve airline operations.

Dana Dabiri and Charles Pecora

Chapter 1

Introduction

Turbulent fluid flow occurs in all aspect of our lives, both in nature and within our engineered society. At the very least, it affects the atmosphere and air that we and nature breathe; the oceans, seas, rivers, and therefore our natural environment; the heat transfer within it that allows for life; it is what is responsible for allowing airplanes to fly and boats and ships to navigate the oceans; when combined with chemical reactions, it is responsible for the power generation and its consequential pressure and temperature distributions within our vehicles' (i.e., cars, ships, airplanes) engines; and it is responsible for the oxygen, heat, and pressure transport when we breath air into our lungs, as well as the blood flow within our hearts and our cardiovascular system, and its necessary pressure distributions that keep us alive and healthy. Because of its all-pervasiveness in all matters of life, understanding turbulent flow is paramount for the progress of our society, and therefore its accurate measurement is critical to attaining this achievement.

Historically, flow visualization has been the primary technique for analyzing turbulence. As early as the 1400s, Leonardo da Vinci recognized that flow needed to be visualized and used fine particles to trace turbulent flows in the aortic heart valve [1]. Ludwig Prandtl observed flow separation over an airfoil and other two-dimensional bodies using floating mica particles in a water tunnel [2]. Fage and Townend [3] quantified flow visualization by capturing streak photographs of fluid flow and manually measuring fluid motion [4]. Pickering and Halliwell, using laser speckle photography (LSV) methodologies, heavily seeded the flow with reflective particles such that individual particles could not be distinguished, illuminated a cross-section of the flow of interest with a laser sheet, and photographically recorded sequences of images of the flow evolution. Using LSV methodologies, they were able to quantify the kinematic flow field, thereby initiating the field of automated quantitative particle velocimetry [5–8]. Shortly thereafter, reduced seeding allowed for imaging of individual particles, giving rise to particle image velocimetry (PIV) and particle tracking velocimetry (PTV).

doi:10.1088/978-0-7503-2203-4ch1

PIV techniques do not match individual particles to determine flow velocities. Instead, an interrogation window subsamples an image pair, which is analyzed with a spatial correlation method, resulting in a single velocity vector that represents the average velocity of particles within the interrogation window. The interrogation window is then systematically moved through the image pair, yielding an evenly spaced velocity vector field. This averaged result introduces smoothing and bias errors in the presence of sharp flow gradients, such as in boundary layers and in shockwaves, thereby reducing the accuracy of the measurements [9, 10]. In addition, the size of this correlation window limits the spatial resolution of PIV, as the interrogation windows are typically not overlapped more than 50%.

The field of PTV became prevalent when Adamczyk and Rimai [11] introduced an automatic particle tracking algorithm for low particle densities. This improved upon previous work in PTV, which was founded on manual measurements of particle displacements [12]. This early work streak-exposed each frame; however, modern PTV uses singly exposed frames with a pulsed laser, and matches particle pairs from frame to frame. Particle velocities are then found by calculating the displacement between matched particle pairs and dividing by their displacement times.

Naturally, the question arises as to whether the limitations of PIV, as described above, can be overcome by using PTV. Using simulated images of a step-like displacement profile, Kähler *et al* [13] studied the resolution limits of PIV. Characterizing the vector spatial resolution as the width of the response function (step response width (SRW)), they compared PIV processed with 16 × 16 windows, single-pixel ensemble-correlations, and PTV (see figure 1.1), and found that PTV showed an order of magnitude lower SRW compared to PIV. In addition, they showed that the dynamic spatial range (DSR) [14] was almost three orders of magnitude larger than PIV processed with 8 × 8 windows. PTV therefore promises the opportunity to substantially increase the spatial resolution as well as the measurement accuracies.

While PIV has thrived since its development [4, 15–23], PTV has until recently received less attention. It is therefore the goal of this book to provide a review of particle tracking velocimetry, focusing on the methods and techniques that have been implemented towards developing and improving the PTV methodology, rather than on its applications. In this regard, chapter 2 discusses experimental components and their set-up. Chapters 3 through 8 are laid out according to the steps that PTV processing algorithms typically operate: chapter 3 discusses particle identification algorithms within images; chapter 4 discusses algorithms that use these to identify particle locations in real space for both 2D and 3D methodologies; and chapter 5 discusses algorithms on how the particles are tracked between image pairs in both 2D and 3D. Chapter 6 explains methods that do not follow the typical sequential approach used in PTV methodologies. Chapter 7 provides a summary of the 3D methodologies within a table, chapter 8 discusses post-processing techniques and, finally, chapter 9 provides concluding remarks.

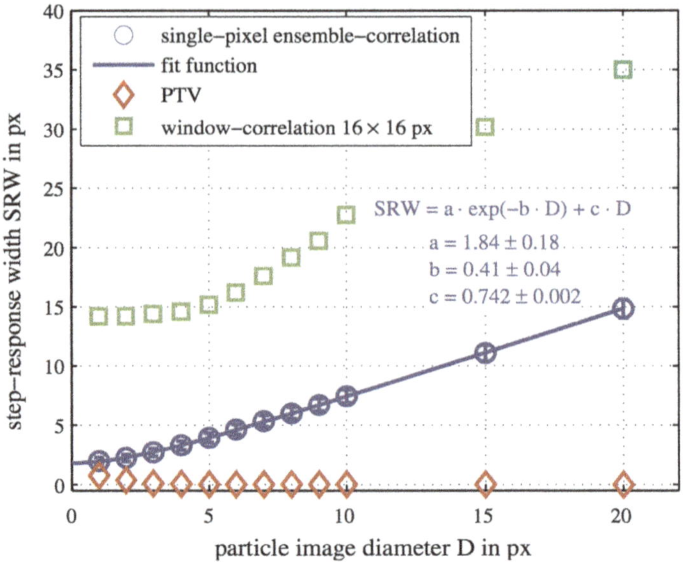

Figure 1.1. Step response width of the estimated displacement with respect to the digital particle image diameter. Reprinted from [13] with permission of Springer.

References

[1] Gharib M, Kremers D, Koochesfahani M M and Kemp M 2002 Leonardo's vision of flow visualization *Exp. Fluids* **33** 219–23

[2] Prandtl L 1904 Über Flüssigkeitsbewegung bei sehr kleiner Reilbung *Proc. Verhandlungen des III. Int. Mathematiker-Kongresses, Heidelberg* (Leipzig: Teubner) pp 404–91

[3] Fage A and Townend H C H 1932 An examination of turbulent flow with an ultra-microscope *Proc. R. Soc. Lond.* A **135** 656–77

[4] Adrian R J 1991 Particle-imaging techniques for experimental fluid mechanics *Annu. Rev. Fluid Mech.* **23** 261–304

[5] Dudderar T D and Simpkins P G 1977 Laser speckle photography in a fluid medium *Nature* **270** 45–7

[6] Simpkins P G and Dudderar T D 1978 Laser speckle measurements of transient benard convection *J. Fluid Mech.* **89** 655

[7] Meynart R 1983 Instantaneous velocity-field measurements in unsteady gas-flow by speckle velocimetry *Appl. Opt.* **22** 535–40

[8] Pickering C J D and Halliwell N A 1984 Speckle photography in fluid-flows—signal recovery with 2-step processing *Appl. Opt.* **23** 1128–9

[9] Scarano F 2003 Theory of non-isotropic spatial resolution in PIV *Exp. Fluids* **35** 268–77

[10] Noguiera J, Lecuona A and Rodriguez R A 2003 Limits on the resolution of correlation PIV iterative methods. Fundamentals *Exp. Fluids* **39** 305–13

[11] Adamczyk A A and Rimai L 1988 2-dimensional particle tracking velocimetry (PTV): technique and image processing algorithm *Exp. Fluids* **6** 373

[12] Dimotakis P E, Debussy F D and Koochesfahani M M 1981 Particle streak velocity field measurements in a two-dimensional mixing layer *Phys. Fluids* **24** 995

[13] Kähler C J, Scharnowski S and Cierpka C 2012 On the resolution limit of digital particle image velocimetry *Exp. Fluids* **52** 1629–39

[14] Adrian R J 1997 Dynamic ranges of velocity and spatial resolution of particle image velocimetry *Meas. Sci. Technol.* **8** 1393

[15] Buchhave P 1992 Particle image velocimetry—status and trends *Exp. Therm Fluid Sci.* **5** 586–604

[16] Grant I 1997 Particle image velocimetry: a review *Proc. Inst. Mech. Eng.* C **211** 55–76

[17] Raffel M, Willert C and Kompenhans J 1998 *Particle Image Velocimetry: A Practical Guide* (Berlin: Springer)

[18] Gharib M and Dabiri D 2000 *An Overview of Digital Particle Image Velocimetry in Flow Visualization: Techniques and Examples* ed A Smits and T T Lim (London: Imperial College Press)

[19] Adrian R J 2005 Twenty years of particle image velocimetry *Exp. Fluids* **39** 1–54

[20] Dabiri D 2006 Cross-correlation digital particle image velocimetry—a review *Turbulencia* ed A S Freire, A liha and B Breidenthal (Rio de Janeiro: ABCM) pp 155–99

[21] Adrian R J and Westerweel J 2011 *Particle Image Velocimetry* (New York: Cambridge University Press)

[22] Gao Q, Wang H P and Shen G X 2013 Review on development of volumetric particle image velocimetry *Chin. Sci. Bull.* **58** 4541–56

[23] Westerweel J, Elsinga G E and Adrian R J 2013 Particle image velocimetry for complex and turbulent flows *Annu. Rev. Fluid Mech.* **45** 409

Chapter 2

Experimental set-up

Figures 2.1 and 2.2 show typical set-ups for 2D and 3D PTV systems, respectively, and as can be seen, they are not different from the set-ups used for 2D and 3D PIV systems. In both cases, the flow is first seeded with particles that act as flow tracers. A pulsed laser light source and appropriate optics are used to create either a thin light sheet for 2D-PTV, or a light volume for 3D-PTV. For 2D set-ups, a charged couple device (CCD) or complementary metal-oxide semiconductor (CMOS) video camera, positioned perpendicular to the light sheet, is synchronized with the pulsed laser in order to capture the positions of the tracer particles at discrete points in time. For 3D set-ups, multiple CCD or CMOS video cameras are situated about the volume of interest, which are also synchronized with the illuminating laser to capture tracer particle images. A data acquisition system records these images from the camera(s) in real-time. We refer the reader to PIV reviews that detail the specifics of these experimental components [1–3, 16–21].

Time-resolved PTV experiments track fluid particles for multiple consecutive time steps in order to analyze transient flow phenomena. Standard CCDs and Nd:YAG lasers, however, limit data acquisition to a rate of 30 Hz, which is far lower than the frequency at which turbulent flow fluctuations typically occur. The recent development of high-speed CMOS cameras and Nd:YLF lasers have made time-resolved measurements on the order of kilohertz possible [5]. Recently, time-resolved PTV systems have become capable of making measurements up to 400 kHz. Although image quality and resolution are compromised for speed, these systems allow for the acquisition of flow spectra at frequencies that are an order of magnitude larger than what hot-wires and Laser Doppler Velocimetry (LDV) can provide [6, 7].

In this section, we discuss experimental set-up features that are relevant to PTV. Section 2.1 discusses tracer particles and their response times, as this is directly relevant to PTV velocity measurements. Sections 2.2–2.4 discuss the interrogation region's illumination and its associated optics, and image acquisition cameras. As these latter components are common to PIV, they are briefly reviewed, and the reader is referred to PIV literature for further details on these topics.

doi:10.1088/978-0-7503-2203-4ch2

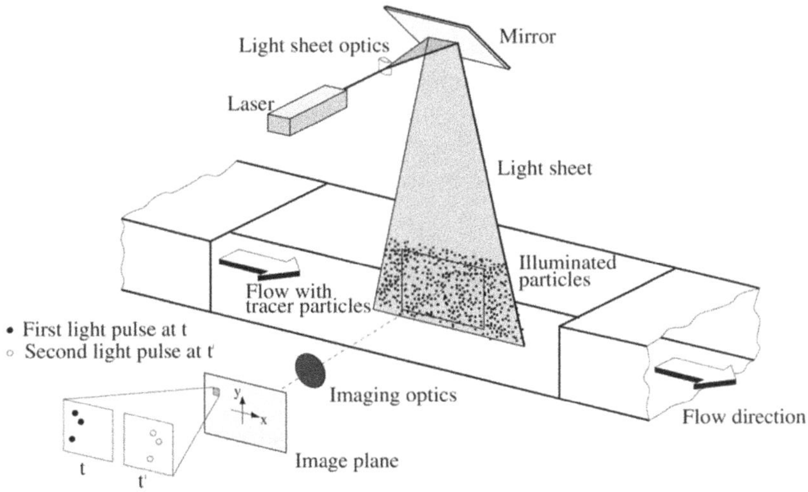

Figure 2.1. Typical layout of a 2D-PTV system for use in a wind or water tunnel. Reprinted from [4] with permission of Springer.

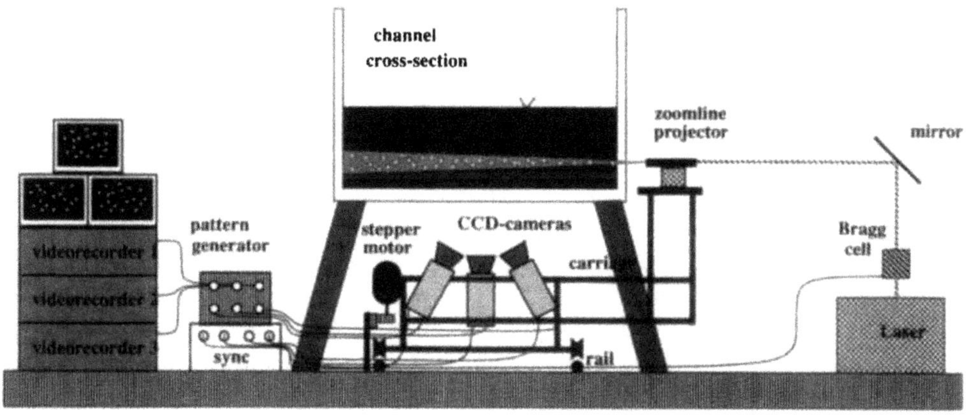

Figure 2.2. Typical layout of a 3D-PTV system for use in a water tunnel. Reprinted from [8] with permission of Springer.

2.1 Tracer particles

In their review and analysis of tracer particles for particle velocimetry, Melling [9] states that a good tracer particle achieves two goals: it follows the local fluid motion and scatters sufficient light for imaging. These requirements are a tradeoff, as smaller particles accurately follow fluid motion with a higher frequency response, while larger particles scatter more light for imaging. An excellent discussion of tracer particle analysis and selection can be found in chapter 2 of *Particle Image Velocimetry* [5], which is clearly relevant to PTV as well. In this section, we summarize the considerations for selecting appropriate tracer particles to achieve accurate flow tracing.

The most basic approach for characterizing a particle response time is by assessing its response to a step function change in fluid velocity. Given that PTV tracer particles' diameters and their velocity lag to such sudden velocity changes are small, this suggests that the particle Reynolds number (Re_p), based on the difference between the flow and the particle velocity (i.e. particle slip velocity), is also small, and that the particle drag is due only to viscous forces. In this case, the particle motion equation is

$$\frac{dU_p}{dt} = K(U - U_p), \tag{2.1}$$

$$\mathrm{Re}_p = (U - U_p)d_p/\nu_f, \tag{2.2}$$

where U is the steady gas velocity, U_p is the particle velocity, and $K = \frac{3C_D\mathrm{Re}_p\mu_f}{4\rho_p d_p{}^2}$ is the entrainment coefficient, which from Stoke's law is given as $18\mu_f/\rho_p d_p^2$. Here, μ_f and ν_f are the fluid dynamic and kinematic viscosity, respectively, and ρ_p and d_p are the particle's density and diameter, respectively. Given that the particle is not in motion at $t = 0$, this equation integrates to

$$U_p(t) = U[1 - e^{-(t/t_s)}], \tag{2.3}$$

where t_s is the particle relaxation time and is given by

$$t_s = \frac{\rho_p d_p^2}{18\mu_f}. \tag{2.4}$$

The particle relaxation time should be less than the smallest time scale of the flow in order to accurately represent the local fluid velocity [10].

Particle response to unsteady flow (i.e. turbulence), however, is not as straightforward. In this case, Mei [11] modified the Maxey–Riley equation for low particle Reynolds number considerations for both solid particles and bubbles. To assess the response of the particles to the high-frequency turbulent fluctuations, the velocity was allowed to oscillate as

$$U(t) = \tilde{U}(\omega)e^{-i\omega t}, \tag{2.5}$$

where $\tilde{U}(\omega)$ is the oscillation amplitude, and the particle Reynolds number is now defined as $\mathrm{Re}_p = \tilde{U}d_p/\nu_f$. Using equation (2.5), Mei expressed the modified Maxey–Riley equation in terms of frequency, and derived an energy response function for the particle that was a function of both the particle Stokes number, $\varepsilon = [(\omega d_p^2)/(8\nu_f)]^{0.5}$, and particle-to-fluid density ratio, $\rho = \rho_p/\rho_f$. [12]. Then, defining cut-off frequencies of the particle based on either 50% or 200% of the energy response, Mei derived approximations for the high-frequency cut-off for density ratios greater than one (solids) and less than one (bubbles). For example, for $\rho < 0.561$, the cut-off frequency is

$$f_{\text{cut-off}} \approx \frac{4\nu_f}{\pi d_p^2}\left[2.38^{0.93} + \left(\frac{0.659}{0.561 - \rho} - 1.175\right)^{0.93}\right]^{2/0.93}, \tag{2.6}$$

while for $\rho > 1.621$, the cut-off frequency is

$$f_{\text{cut-off}} \approx \frac{4\nu_f}{\pi d_p^2}\left[\frac{3}{2\rho^{0.5}}^{1.05} + \left(\frac{0.932}{\rho - 1.621}\right)^{1.05}\right]^{2/1.05}, \tag{2.7}$$

where from the definition of the Stokes number, the high-frequency cut-off can be related to the Stoke's cut-off value

$$\varepsilon_{\text{cut-off}} \approx \frac{d_p}{2}\sqrt{\frac{\pi f_{\text{cut-off}}}{\nu_f}}. \tag{2.8}$$

Then by combining the appropriate estimate for the high-frequency cut-off from equations (2.6) and (2.7) with equation (2.8), one can find the largest particle diameter that would that would respond to the unsteady fluid flow.

For cases in which the density ratio is between these two ranges, the energy transfer function is shown to be less than two, thereby implying a very good particle response. For neutrally buoyant particles, $\rho = 1$, the energy transfer function is equal to one, implying perfect particle response. In further evaluations of the particle response to isotropic turbulence, turbulence energy and Taylor micro-scale structures, Mei concluded that the particle inertia parameter (the inverse of the particle response time) defined as

$$\beta = \frac{9}{2}\frac{\nu}{(\rho + 1/2)a^2}, \tag{2.9}$$

should be large for solid particles in gas flow, and also large with $\rho \approx 1$ for solid particles in liquids.

By further considering the motion of particles in turbulent flow, Melling [9] derived the energy response function for the simplified case of high particle-to-fluid density ratios and low particle Reynolds number, and showed that the ratio of fluctuation intensities of particle and fluid motions is

$$\frac{\bar{u}_p^2}{\bar{u}_f^2} = (1 + t_s\omega_c)^{-1}, \tag{2.10}$$

where ω_c is the highest turbulent frequency of interest, and $\bar{u}_p^2$ and $\bar{u}_f^2$ are the particle and fluid velocity fluctuations, respectively. Lower relaxation times allow for better flow tracking by a particle. For example, a relaxation time of 8.3 μs roughly corresponds to a 1 μm water droplet in air.

These previous works and others [13–16] have suggested that a low particle Stokes number, defined as the ratio of the tracer relaxation time to the Kolmogorov time scale of the flow, would allow for a tracer to reliably follow fluid motion. Recently, however, Mathai *et al* [17] suggested that even for sufficiently small Stokes numbers, the density ratio can cause an increase of acceleration variance. To understand this, they derive the non-dimensional particle equation of motion,

$$\ddot{\boldsymbol{x}} = \beta\frac{D\boldsymbol{u}}{Dt} + \frac{1}{\mathrm{St}}(\boldsymbol{u} - \dot{\boldsymbol{x}}) + \frac{1}{\mathrm{Fr}}\hat{e}_z, \tag{2.11}$$

$$\mathrm{St} \equiv \frac{d_p^2}{12\beta\nu\tau_n}, \tag{2.12}$$

$$\mathrm{Fr} \equiv \frac{a_n}{(\beta-1)g}, \tag{2.13}$$

where $\ddot{\boldsymbol{x}}$ is the acceleration of a tracer particle, β is the density ratio defined as $3\rho_f/(\rho_f + 2\rho_p)$, $D\boldsymbol{u}/Dt$ is the fluid acceleration, St is the Stokes number, Fr is the Froude number, $\boldsymbol{u}$ is the fluid velocity, $\dot{\boldsymbol{x}}$ is the particle velocity, d_p is the particle diameter, ν is the fluid kinematic viscosity, τ_η is the Kolmogorov time scale, a_η is the turbulent acceleration, and g is the gravitational constant. Note that all dimensionless numbers and the fluid acceleration term are adjusted with the density ratio, β. They therefore suggest that an additional constraint for tracer selection should be that the ratio of the Stokes to the Froude number should be less than unity. This constraint is particularly relevant for flows with small Fr, as this indicates a significant contribution from gravitational acceleration.

2.1.1 Tracers in liquid flows

As can be concluded from the previous section, one way to ensure fast response times of tracer particles is to have a particle density nearly the same as the fluid. In liquid applications, a near unity density ratio is achievable, and thus tracer sizes can be increased to improve light scattering. For this reason, less powerful illumination sources can be used for liquid than for gas in experiments. An overview of tracers that have been documented for liquid experiments is shown in table 2.1.

Fluorescent-dyed particles, which emit a different wavelength of light than they absorb, are commonly used in micro-PTV and liquid flows for the purpose of removing reflected light and noise from images. They are typically used with optical filters, which can isolate light fluoresced by fluorescent particles from light reflected by surfaces. Two commonly used fluorescent dyes are Fluorescein and Rhodamine 6G [5, 18]. Fluorescent-dyed particles have also been used for particle velocimetry measurements near walls to allow for improved laser flare removal from acquired images [19, 20]. KR620-doped polystyrene microspheres, for example, have demonstrated the ability to effectively reduce laser reflections, allowing for near-wall measurements, as shown in figure 2.3.

According to Melling, particle seeding in liquid flows is typically done by simply mixing solid particles into the flow. Overbruggen *et al* [21] suggest that in some liquid flows, solid particle sedimentation can be an issue, and thus they introduced a gaseous bubble seeding technique. The technique generated air bubbles in water with

Table 2.1. Seeding particles used in liquid flows [9].

Material	d_p (μm)	Laser	CW power or energy, time	Light sheet w (mm)	t (mm)	Reference
TiO_2	3	Nd:YAG				Longmire and Alahyari (1994)
Al_2O_3	9.5	Ruby	2 J, 30 ns	100	0.8	Liu *et al* (1991)
Conifer pollen ($\rho = 1000$ kg m^{-3})	50–60	Ar ion	1–2 W			Westergaard *et al* (1993) McCluskey *et al* (1995) Gallagher and McEwan (1996)
Polymer ($\rho = 1030$ kg m^{-3})	30	Ar ion	0.5–5 W		0.5	Draad and Westerweel (1996) McCluskey *et al* (1996)
Phosphorescent polymer	80	Ar ion	5 W		1	Willert and Gharib (1991)
Fluorescent	50	Nd:YAG				Hart (1996)
	20	Cu vapor	45 W		1	Roth *et al* (1995)
Polystyrene	500					Khoo *et al* (1992)
($\rho = 1050$ kg m^{-3})	15	Ruby	25 mJ, 20 ns			Zhang *et al* (1996)
Thermoplastic ($\rho = 1020$ kg m^{-3})	6	Nd:YAG		50	2	Hassan *et al* (1994)
Reflective	60	Ar ion	18 W			Grant *et al* (1992)
($\rho = 1010$ kg m^{-3})	30	Ar ion	12–18 W	200		Grant and Wang (1994)
Metallic coated	4	Ar ion	2 W		2	Magness *et al* (1993)
	14	Ar ion			1	Johari *et al* (1996)
Microsperes ($\rho = 700$ kg m^{-3})	<30	Ar ion				Graham and Soria (1994)
H_2 bubbles		Ar ion	1 W		0.3	Dieter *et al* (1994)

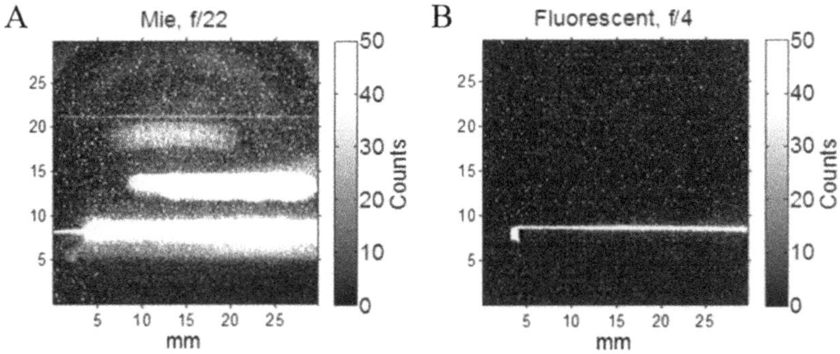

Figure 2.3. Raw PIV images with a flat plate for Mie-scattering (A) and fluorescent light (B). Reprinted with permission from [19]. Copyright (2015) American Chemical Society.

a median diameter of 10.64 μm and a standard deviation of 5.3 μm. Despite the low-density ratio of the bubbles to water, the relaxation time is estimated to be 0.629 ns and the buoyancy effects were assumed to be negligible.

2.1.2 Tracers in subsonic gas flows

Tracer particles in gas flow typically have a density ratio much greater than 1. For a given gas and tracer particle density, the particle diameter must be selected to achieve a sufficiently small relaxation time. Proper generation of seeding particles is therefore important [9], and is typically done with a Laskin nozzle [22]. Table 2.2 shows that for a 1 kHz response, 2–3 μm diameter tracer particles suffice, while a 10 kHz response requires less than 1 μm diameter particles; these frequency limits were calculated using the simplified frequency response for high particle-to-fluid density ratios given in equation (2.9) [9]. Some applications of particle imaging in gas flows to various experiments are listed in table 2.3, showing that they generally adhere to the calculated maximum diameters.

In gas flows, liquid droplets are typically introduced using a Laskin nozzle, which reproducibly creates droplets on the order of 1 μm at concentrations as high as 1–10 mm^{-3}. Commercially available smoke generators, on the other hand, create a water-based or oil-based smoke using the principle of evaporation followed by condensation. Although less consistent, these smoke generators can create higher particle concentrations and smaller tracer particles. Solid particles are commonly seeded via atomization, creating an aerosol [9].

Helium-filled soap bubbles (HFSB) have recently been implemented as tracer particles in wind tunnel experiments [23–25]. Unlike the micron-scale tracer particles typically used, these soap bubbles are estimated to be around 400 μm and neutrally buoyant. The larger tracers scatter light with intensity several orders of magnitude higher than micron sized droplets. Due to this increased brightness, the HFSB

Table 2.2. Particle response in turbulent flow (η = 0.99) [9].

Particle	ρ_p (kg m^{-3})	Gas (10^5 Pa)	Density ratio s	Viscosity ν (m^2 s^{-1})	f_c (kHz)	Sk_c	d_p (μm)
TiO_2	3500	Air	2950	1.50×10^{-5}	1	0.0295	1.44
		(300 K)			10		0.45
Al_2O_3	3970	Flame	20250	3.00×10^{-4}	1	0.0113	2.46
		(1800 K)			10		0.78
Glass	2600	Air	2190	1.50×10^{-5}	1	0.0342	1.67
		(300 K)			10		0.53
Olive oil	970	Air	617	1.45×10^{-5}	1	0.0645	3.09
		(220 K)			10		0.98
Microballoon	100	Air	84.5	1.50×10^{-5}	1	0.1742	8.50
		(300 K)			10		2.69

Table 2.3. Seeding particles used in gas flows [9].

Material	d_p (μm)	Laser	Pulse energy, pulse time	Light sheet w (mm)	Light sheet t (mm)	Reference
TiO_2 ($m = 2.6$, $\rho = 3500$ kg m^{-3})	<1	Nd:YAG	10 mJ, 20 ns	15	0.3	Reuss *et al* (1989)
TiO_2 ZrO_2	0.7 − 1	Nd:YAG	110 mJ, 12 ns			Paone *et al* (1996)
Al_2O_3 ($m = 1.76$,	0.3	Nd:YAG	400 mJ		0.2	Muniz *et al* (1996)
$\rho = 3500$ kg m^{-3})	3	Nd:YAG	9 mJ, 6 ns			Anderson *et al* (1996)
	0.8	Ruby	20 ns	150	~1	Krothapalli *et al* (1996)
Polycrystalline	30	Nd:YAG	135 mJ, 6 ns			Grant *et al* (1994)
Glass	30	Ruby	30 mJ, 30 ns			Schmidt and Löffler (1993)
Oil smoke	1	Ruby	5 J			Stewart *et al* (1996)
Corn oil	1 − 2	Nd:YAG	100 mJ			Jakobsen *et al* (1994)
Oil	1 − 2	Nd:YAG	120 mJ		0.4	Westerweel *et al* (1993)
Olive oil	1.06	Nd:YAG	70 mJ, 16 ns	200	0.5	Höcker and Kompenhans (1991)
($m = 1.47$,						Fischer (1994)
$\rho = 970$ kg m^{-3})						Raffel *et al* (1996)

tracers are useful in PTV experiments with wind tunnels and measurement volumes that are two orders of magnitude larger than that of a conventional experiment. HFSB generation is achieved by feeding coaxial channels with constant flow rates of helium, bubble fluid solution (water, glycerin, and soap), and air. The coaxial channels terminate with a small circular orifice. The flow rates of helium and bubble fluid can be adjusted so as to match the density of air for a minimum response time. The minimum characteristic response time was been reported to be 10 μs [21].

2.1.3 Tracers in supersonic flows

In transonic and supersonic flows, particle diameters must be further reduced, as shockwaves create substantial instantaneous changes in velocity [10], thereby requiring faster particle response times for accurate flow determination. Seeding particles for supersonic flows are typically liquids, which can be seeded using a Laskin nozzle [22]. Using equations (2.1) through (2.4), Chen and Emrich [10] estimated the time required for a tracer particle at rest in a stream of velocity u to acquire the fraction v/u (v is the particle velocity) of the final speed in a shock tube with a pressure ratio of 2.0 in figure 2.4.

Melling [9] recognized that the expression for the drag coefficient needed to be modified for compressibility effects in transonic and supersonic flow, and therefore proposed the modified Stokes drag, by incorporating the particle Knudsen number,

Kn_p, as shown in equation (2.14). This was then used in the entrainment coefficient within a two-dimensional form of equation (2.1), to obtain the particle relaxation times:

$$C_D = \frac{24}{\mathrm{Re}_p(1 + \mathrm{Kn}_p)}. \tag{2.14}$$

Tedeschi *et al* [26] proposed a model for the drag coefficient of tracer particles that takes into account compressibility and rarefaction effects in all rarefaction regimes (Kn_p) when $M \leqslant 1$ and $\mathrm{Re}_p \leqslant 200$:

$$C_D = \frac{24}{\mathrm{Re}_p} k\left[1 + 0.15(k\mathrm{Re}_p)^{0.687}\right]\xi(\mathrm{Kn}_p)C, \tag{2.15}$$

where

$$k = \left(1 + \frac{9}{2}\mathrm{Kn}_p\right), \tag{2.16}$$

$$C = 1 + \frac{\mathrm{Re}_p^2}{\mathrm{Re}_p^2 + 100} e^{-0.225/M^{2.5}}, \tag{2.17}$$

$$\xi(\mathrm{Kn}) = 1.177 + 0.177\frac{0.851\mathrm{Kn}_p^{1.16} - 1}{0.851\mathrm{Kn}_p^{1.16} + 1}, \tag{2.18}$$

and M is the relative Mach number between the particle and the fluid. This expression is an extension of Cunningham's [27] method to higher velocities, and was shown to agree with experimental results across an oblique shockwave.

Loth [28] introduced a drag model for tracer particles that divides the problem into a compression-dominated ($\mathrm{Re} \leqslant 45$) and a rarefaction-dominated ($\mathrm{Re} \geqslant 45$) regime. Within the compression-dominated regime, the drag coefficient is

$$C_D = \frac{24}{\mathrm{Re}_s}\left[1 + 0.15\mathrm{Re}_s^{0.687}\right]H_M + \frac{0.42C_M}{1 + \frac{4200G_M}{\mathrm{Re}_s^{1.16}}}, \tag{2.19}$$

$$H_M = 1 - \frac{0.258C_M}{1 + 514G_M}, \tag{2.20}$$

$$G_M = \begin{array}{ll} 1 - 1.525Ma_s^4 & Ma_s < 0.89 \\ 0.0002 + 0.0008\tanh[12.77(Ma_s - 2.02)] & Ma_s > 0.89 \end{array}, \tag{2.21}$$

$$C_M = \begin{array}{ll} \frac{5}{3} + 2/3\tanh\left[3\ln(Ma_s/1.5)^2\right] & Ma_s < 1.45 \\ 2.044 + 0.2\exp\left[-1.8\left[\ln(Ma_s/1.5)\right]^2\right] & Ma_s > 1.45 \end{array}. \tag{2.22}$$

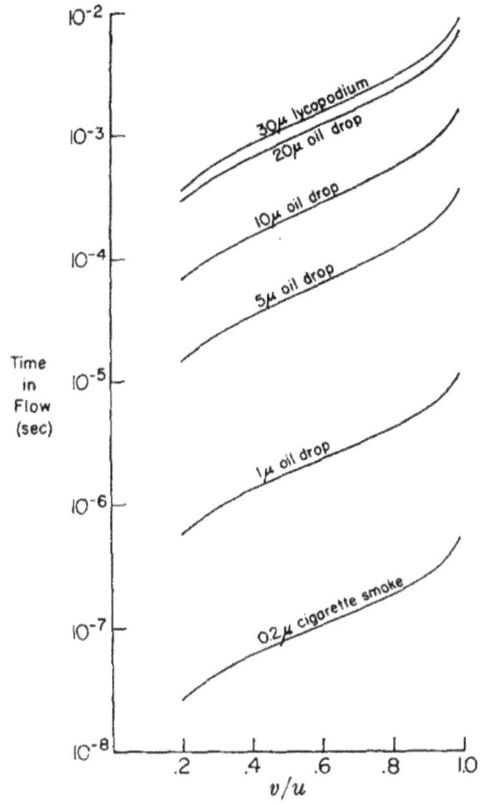

Figure 2.4. Step response of tracers of various sizes to a shockwave. Reprinted from [10], with the permission of AIP Publishing.

For drag within the rarefaction-dominated regime:

$$C_D = \frac{C_{D,\mathrm{Kn,Re}}}{1 + M_s^4} + \frac{M_s^4 C_{D,fm,\mathrm{Re}}}{1 + M_s^4}, \tag{2.23}$$

$$C_{D,\mathrm{Kn,Re}} = \frac{24}{\mathrm{Re}_s}\left(1 + 0.15R_s^{0.687}\right) f_{\mathrm{Kn}}, \tag{2.24}$$

$$f_{\mathrm{Kn}} = \frac{1}{1 + \mathrm{Kn}[2.514 + 0.8\exp(-0.55/\mathrm{Kn})]}, \tag{2.25}$$

$$C_{D,fm,\mathrm{Re}} = \frac{C_{D,fm}}{1 + \left(\frac{C_{D,fm}}{1.63} - 1\right)\sqrt{\frac{\mathrm{Re}_s}{45}}}, \tag{2.26}$$

$$C_{D,fm} = \frac{(1 + 2s^2)\mathrm{erf}(-s^2)}{s^3\sqrt{\pi}} + \frac{(4s^3 + 4s^2 - 1)\mathrm{erf}\,(s)}{2s^4} + \frac{2}{3s}\sqrt{\pi}, \tag{2.27}$$

$$s \equiv M_s\sqrt{\gamma/2}\,. \tag{2.28}$$

The two regimes reach a nexus point of Re = 45, where the drag coefficient is equal to 1.63, and is independent of Mach number and Knudsen number. This model was created using a wide range of experimental data and proved to better predict tracer response compared to previous models. The model correlated well with experimental data; however, it does not incorporate unsteadiness or non-sphericity effects.

Most recently, Williams *et al* [29] suggest that particle response tests done across a strong shockwave tend to overestimate the response of particles to turbulence. It was also shown that rarefaction effects reduce the frequency response of particles in turbulent high-speed flows. Thus, in order to have confidence in tracer selection for supersonic flows, for weakly turbulent flow, they recommend iteratively solving for the particle size and density, and then using the Cunningham [27] slip correction, C_c, to correct the Stokes response frequency:

$$\frac{f}{f_s} = \frac{1}{C_c}, \tag{2.29}$$

$$C_c = 1 + \mathrm{Kn}\left(A_1 + A_2\exp\left(\frac{-A_3}{\mathrm{Kn}}\right)\right), \tag{2.30}$$

where the constants are given as $A_1 = 2.541$, $A_2 = 0.8$, and $A_3 = 0.55$. For strong turbulence, they recommend performing a full simulation using the Loth drag model to obtain the particle drag coefficient.

For time-resolved supersonic flow systems, droplets have been generated with diameters between 0.6 and 0.7 μm using an oil-based smoke generator, where their highest Stokes number have been estimated *a posteriori* to be 0.05, and were therefore assessed to be sufficiently small to neglect particle lag errors [6].

Ghaemi *et al* [30] investigated the use of aluminum nanoparticles with diameters around 10 nm by imaging the tracers in a 3×3 cm^{-2} supersonic wind tunnel. The particles were imaged in an oblique shockwave at Mach 2. The aluminum nano-structured aggregates were generated by the spark discharge mechanism, in which two opposing cylindrical electrodes are located a few millimeters apart. The electrodes were continuously charged, until a breakdown voltage was reached, at which point a spark discharge vaporizes the electrode material. The material rapidly cooled, generating a high concentration of nanoparticles. The velocity normal to the shockwave was measured using PIV for these nanoparticles as well as commonly used TiO_2 particles with a primary diameter of 50 nm. The shockwave was located at 0 mm in figure 2.5. The aluminum nanoparticles reached the downstream velocity after traveling 0.3 mm, while the TiO_2 particles required about 2 mm. The corresponding relaxation times were determined to be 0.27 and 2.0 μs for aluminum and TiO_2 particles, respectively.

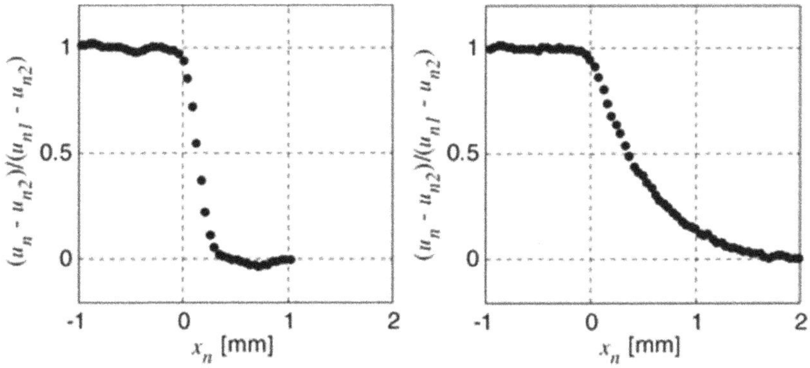

Figure 2.5. PIV measurement of the normal velocity of aluminum nanoparticles (left) and TiO_2 particles (right) across an oblique shockwave located at 0 mm [30].

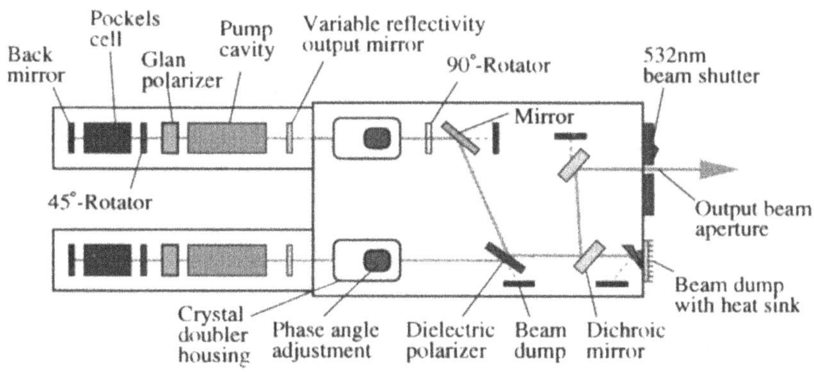

Figure 2.6. Dual-cavity Nd:YAG laser. Reprinted from [31] with permission of Springer.

2.2 Illumination

Modern PTV systems, like PIV systems, require a pulsed light source with a short pulse duration and fast pulse frequency. The most commonly used lasers are Nd: YAG, which emit infrared radiation at 1064 nm and that are doubled to 532 nm, allowing for pulsed visible illumination. The lasers can typically deliver between 12 and 1000 mJ per pulse at 10 to 30 Hz[5]. The laser pulse frequencies are synchronized with the video camera frame rate to ensure singly exposed images by exposing the first image of the pair at the end of its frame, and exposing the second image of the pair as the beginning of its frame. This is a process commonly called 'frame-straddling'. To ensure this type of asynchronous laser pulsing of the image pairs, PIV/PTV lasers are two Nd:YAG lasers housed together, where their beams are overlapped and combined into a single beam (see figure 2.6).

Recent technological advances have allowed for the development of the Nd:YLF laser, which outputs for 10–20 mJ per pulse at 527 nm at rates up to 10 kHz. These

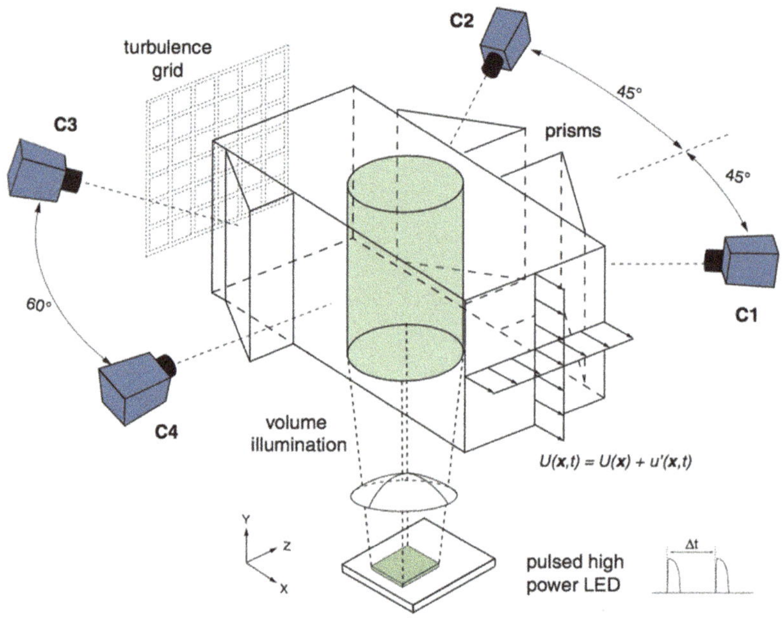

Figure 2.7. LED illumination for tomographic PIV. Reprinted from [36] with permission of Springer.

lasers are also dual-cavity and share the same set-up as Nd:YAG lasers shown in figure 2.6. When combined and synchronized with high-speed cameras, these lasers have allowed for time-resolved (TR) studies, which allow for high temporal resolution data acquisition that captures the time evolution of turbulent flows [32].

An emerging technology for volume illumination is provided by pulsed LED arrays, which provide 1–10 mJ of energy per pulse and pulse rates in excess of 2 kHz [33–36]. The benefits of using LED illumination for PTV are that they are a low cost and easy to use alternative to lasers. Additionally, LEDs can be bundled together into compact arrays to increase the illumination volume size or intensity [36]. A 3D PIV schematic set-up using LED volume illumination is shown in figure 2.7. Also, figure 2.8 shows a schematic set up of an experiment that combines LED and laser illumination for recording of particle images in bubbly flow. The laser illuminates fluorescent tracer particles within a 2D light sheet. Reflected laser light is filtered and the LED backlight creates shadow images of the bubbles. This set-up results in a single camera recording containing both particle and bubble images.

2.3 Area/volume illumination optics

The purpose of the laser light source is to illuminate a desired area or volume of interest. A series of lenses is typically arranged to manipulate the light into a sheet or volume. Figure 2.9 shows an example configuration that uses three cylindrical lenses for laser sheet generation. The first diverging lens stretches the width of the laser

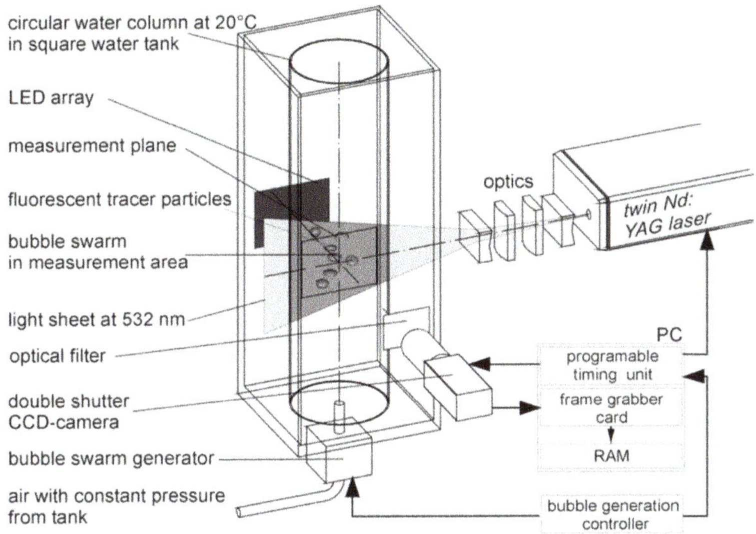

Figure 2.8. Experimental set-up using laser and LED illumination for investigation of bubbly flow. Reprinted from [33] with permission of Springer.

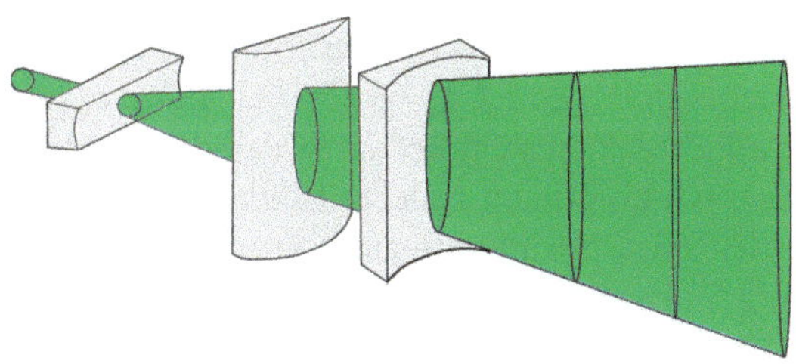

Figure 2.9. Example optics for light sheet generation adapted from [37].

sheet. The following converging and diverging lenses are rotated 90 degrees with respect to the first lens, where their separation controls the sheet thickness.

A laser can be manipulated to illuminate a volume in a similar manner as shown above. One technique is to use a beam expander with variable magnification. This has been used to illuminate a cylindrical volume with a 30 mm diameter [38, 39]. Alternatively, two diverging cylindrical lenses can be oriented perpendicular to one another in order to expand the laser into an ellipsoidal cone [40]. Another common practice in volume illumination is the use of beam blocks to remove the low intensity edges of the light and sharply define the volume thickness [40, 41].

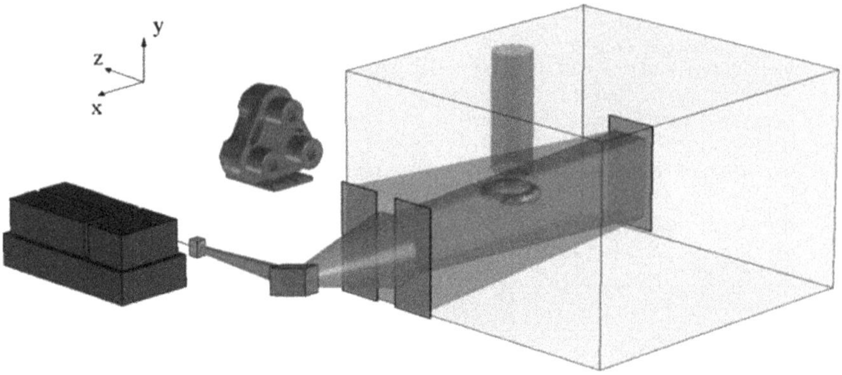

Figure 2.10. The use of dielectric mirrors to increase the light incident on a volume. Reprinted from [40] with permission of Springer.

Due to the high energy required to illuminate a volume sufficiently for PTV, a technique of using two dielectric mirrors to reflect light back and forth through the test section can be used (figure 2.10). This technique has been used in volume illumination to effectively increase the laser pulse energy from 8 mJ to 60 mJ [40–42].

2.4 Camera

When quantitative particle velocimetry methods were initially developed, film cameras were used. Due to their frame rate limitation, frames were typically doubly exposed [5]. With the advent of new technology, however, particle velocimetry image acquisition systems now use CCD or CMOS video cameras [43]. These video cameras have an $m \times n$ light-sensitive pixel array (available from a 1 megapixel array to 50 megapixel array), which collect reflected light from the tracer particles. CCD video cameras typically incorporate interline frame-transfer chips that allow for interframe laser pulse separations as small as 200 ns [5]. The CMOS video cameras incorporate circuitry for each pixel that performs signal processing, which allows for very fast readout. This makes CMOS cameras suitable for high-speed image acquisition as high as 5 kHz. Time-resolved PTV is typically achieved using CMOS cameras and Nd:YLF lasers [5]. An analog-to-digital converter discretizes the analog intensity values read by the camera. A frame grabber that resides on the computer interfaces with the camera to collect data and store it in the computer memory.

In using image recording devices, noise is inherent in each pixel. This noise is random and introduces error in particle identification. Three types of error exist in digital recording: background noise, shot noise, and device noise. Background noise refers to the camera output even in the absence of incident light, typically referred to as 'dark current'. It includes both a mean component, which is typically about 30 counts out of 256 for 8-bit cameras, and a random component that is characterized by a Poisson distribution. Shot noise is the random fluctuation inherent to the conversion of light energy to electron flow. It is independent for each pixel and is also characterized by a Poisson distribution. Device noise is caused by a device's internal electronics, which includes thermal noise, semiconductor surface noise, and

Table 2.4. Comparison of cameras used in PTV, by Hain *et al* [44]. Cameras are labeled with superscripts 'a' (CCD), 'b (CMOS), 'c' (intensified CMOS), 'd' (identical to LaVision Flowmaster 3 s), 'e' (CMOS interframe time), and 'f' (identical to IDT XS5-i). fps refers to frames per second and IF means interframe time.

Camera	Resolution (px^2)	Pixel size (μm^2)	fps (1/s)	IF (ns)
PCO Sensicam[a,d]	1280 × 1024	6.7 × 6.7	8	300
DantecDynamics NanoSense MKIII[b]	1280 × 1024	12 × 12	1024	100
DantecDynamics iNanoSense MK7[c,f]	1280 × 1024	12 × 12	1024	100
PCO 1200hs[b]	1280 × 1024	12 × 12	638	75
Photron APX[b]	1024 × 1024	17 × 17	2000	2000
Photron APX-i2[c]	1024 × 1024	17 × 17	2000	2000[e]
Redlake HG lOOK[b]	1504 × 1128	12 × 12	1000	3000

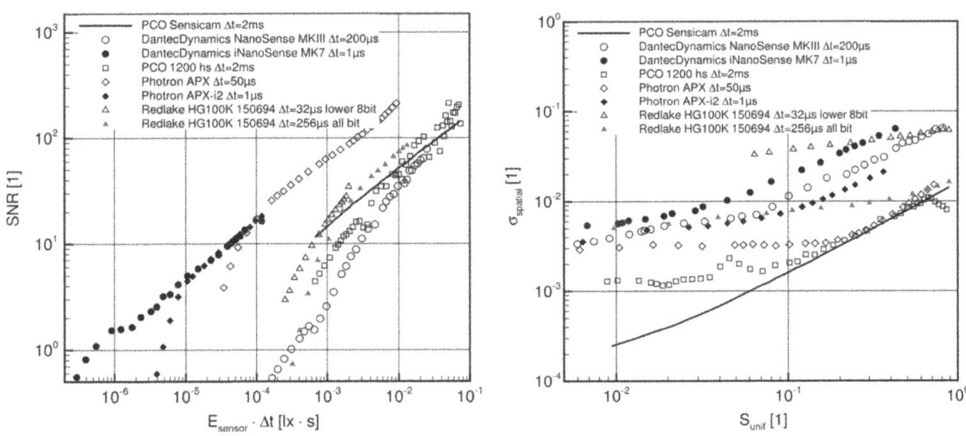

(a) (b)

Figure 2.11. (a) Dependence of SNR on incident light exposure and (b) spatial pixel variance versus the uniform signal level. Reprinted from [44] with permission of Springer.

switching noise. These sources of noise adhere to a Gaussian distribution. For the purposes of PTV, a camera with high signal-to-noise ratio (SNR) is desirable, and is commonly designed by parametric optimization of these sources of noise [5].

Hain *et al* [44] performed a study of digital cameras to quantify and compare the performance of different digital cameras, which can be used to help select an appropriate sensor for a desired PTV application. The CCD, CMOS, and intensified CMOS cameras used are listed in table 2.4. Note that the CMOS cameras' pixels are four to six times larger and frame rates two to three orders of magnitude higher than the CCD.

Figure 2.11(a) shows how the SNR of the tested cameras varies over a range of light exposure values, measured in lux-seconds. In considering SNR, intensified CMOS cameras (iNanoSense MK7 and Photron APX-i2) perform well with low light intensity and are recommended when illumination is limited. The CMOS cameras (NanoSense MKIII, PCO 1200hs, Photron APX, and Redlake HG 100K) generally also have a higher SNR for a given light intensity than the CCD (PCO Sensicam) due to their larger pixel size and fill factor. In addition to the testing of SNR, a uniform signal, S_{unif}, was generated for each light intensity by temporally averaging 20 images. For each of these averaged uniform signals, the spatial standard deviation of pixel gray values was calculated and is shown in figure 2.11(b) as a function of S_{unif}. The CCD has the lowest $\sigma_{spatial}$, and therefore superior image uniformity, in all conditions. The quality of the images can be further improved by CCD cooling. The CMOS cameras perform almost as well as the CCD for highly uniform signals. The intensified CMOS cameras have inferior uniformity for all signal levels. For this reason, Hain recommends that CCDs should be used for the highest precision measurements. CMOS cameras should be used for time-resolved studies, as they are capable of higher frame rates than CCDs. A CMOS with an intensifier is recommended only when illumination levels are otherwise insufficient.

References

[1] Adrian R J 1991 Particle-imaging techniques for experimental fluid mechanics *Annu. Rev. Fluid Mech.* **23** 261–304

[2] Buchhave P 1992 Particle image velocimetry—status and trends *Exp. Therm Fluid Sci.* **5** 586–604

[3] Gao Q, Wang H P and Shen G X 2013 Review on development of volumetric particle image velocimetry *Chin. Sci. Bull.* **58** 4541–56

[4] Raffel M, Willert C and Kompenhans J 2007 *Particle Image Velocimetry: A Practical Guide* 2nd edn (Berlin: Springer)

[5] Adrian R J and Westerweel J 2011 *Particle Image Velocimetry* (New York: Cambridge University Press)

[6] Beresh S J, Henfling J F and Spillers R W Postage-stamp PIV: small velocity fields at 400 kHz for turbulence spectra measurements *AIAA SciTech Forum (Grapevine, Texas, 9–13 January 2017)*

[7] Willert C E 2015 High-speed particle image velocimetry for the efficient measurement of turbulence statistics *Exp. Fluids* **56** 17

[8] Maas H G, Gruen A and Papantoniou D 1993 Particle tracking velocimetry in three-dimensional flows Part 1. Photogrammetric determination of particle coordinates *Exp. Fluids* **15** 133–46

[9] Melling A 1997 Tracer particles and seeding for particle image velocimetry *Meas. Sci. Technol.* **8** 1406–16

[10] Chen C J and Emrich R J 1963 Investigation of the shock-tube boundary layer by a tracer method *Phys. Fluids* **6** 1

[11] Mei R 1994 Flow due to an oscillating sphere and an expression for unsteady drag on the sphere at finite Reynolds number *J. Fluid Mech.* **270** 174

[12] Mei R 1996 Velocity fidelity of flow tracer particles *Exp. Fluids* **22** 1–13
[13] Toschi F and Bodenschatz E 2009 Lagrangian properties of particles in turbulence *Annu. Rev. Fluid Mech.* **41** 375
[14] Calzavarini E, Volk R, Bourgoin M, Leveque E, Pinton J F and Toschi F 2009 Acceleration statistics of finite-sized particles in turbulent flow: the role of Faxén forces *J. Fluid Mech.* **630** 179
[15] Calzavarini E, Kerscher M, Lohse D and Toschi F 2008 Dimensionality and morphology of particle and bubble clusters in turbulent flow *J. Fluid Mech.* **607** 13
[16] Bec J, Biferale L, Boffetta G, Celani A, Cencini M, Lanotte A, Musacchio S and Toschi F 2006 Acceleration statistics of heavy particles in turbulence *J. Fluid Mech.* **550** 349
[17] Mathai V, Calzavarini E, Brons J, Sun C and Lohse D 2016 Microbubbles and microparticles are not faithful tracers of turbulent acceleration *Phys. Rev. Lett.* **117** 024501
[18] Pedocchi F, Martine J E and Garcia M H 2008 Inexpensive fluorescent particles for large-scale experiments using particle image velocimetry *Exp. Fluids* **45** 183–6
[19] Wohl C J, Kiefer J M, Petrosky B J, Tiemsin P I, Lowe K T, Misto P M and Danehy P M 2015 Synthesis of fluorophore-doped polystyrene microspheres: seed material for airflow sensing *ACS Appl. Mater. Interfaces* **7** 20714–25
[20] Petrosky B J, Lowe K T, Danehy P M, Wohl C J and Tiemsin P I 2015 Improvements in laser flare removal for particle image velocimetry using fluorescent dye-doped particles *Meas. Sci. Technol.* **26** 115303
[21] Overbruggen T, Schroder F, Klass M and Schroder W 2014 A particle-image velocimetry tracer generating technique for liquid flows *Meas. Sci. Technol.* **25** 087001
[22] Kähler C J, Sammler B and Kompenhans J 2002 Generation and control of tracer particles for optical flow investigations in air *Exp. Fluids* **33** 736–42
[23] Biwole P J, Yan W, Zhang Y H and Roux J J 2009 A complete 3D particle tracking algorithm and its applications to indoor airflow study *Meas. Sci. Technol.* **20** 11
[24] Scarano F, Ghaemi S, Caridi G C A, Bosbach J, Dierksheide U and Sciacchitano A 2015 On the use of helium-filled soap bubbles for large-scale tomographic PIV in wind tunnel experiments *Exp. Fluids* **56** 42
[25] Schneiders J F G, Caridi G C A, Sciacchitano A and Scarano F 2016 Large-scale volumetric pressure from tomographic PTV with HFSB tracers *Exp. Fluids* **57** 164
[26] Tedeschi G, Gouin H and Elena M 1999 Motion of tracer particles in supersonic flows *Exp. Fluids* **26** 288–96
[27] Cunningham E 1910 On the velocity of steady fall of spherical particles through fluid medium *Proc. R. Soc.* A **83** 357–65
[28] Loth E 2008 Compressibility and rarefaction effects on drag of a spherical particle *AIAA J.* **46** 9
[29] Williams O J H, Nguyen T, Schreyer A M and Smits A J 2015 Particle response analysis for particle image velocimetry in supersonic flows *Phys. Fluids* **27** 076101
[30] Ghaemi S, Schmidt-Ott A and Scarano F 2010 Nanostructured tracers for laser-based diagnostics in high-speed flows *Meas. Sci. Technol.* **21** 105403
[31] Raffel M, Willert C and Kompenhans J 1998 *Particle Image Velocimetry: A Practical Guide* (Berlin: Springer)
[32] Elsinga G E, Scarano F L, Wieneke B and van Oudheusden B W 2006 Tomographic particle image velocimetry *Exp. Fluids* **41** 933–47

[33] Lindken R and Merzkirch W 2002 A novel PIV technique for measurements in multiphase flows and its application to two-phase bubbly flows *Exp. Fluids* **33** 814–25
[34] Bröder D and Sommerfeld M 2007 Planar shadow image velocimetry for analysis of the hydrodynamics in bubbly flows *Meas. Sci. Technol.* **18** 2513–28
[35] Willert C, Stasicki B, Klinner J and Moessner S 2010 Pulsed operation of high-power light emitting diodes for imaging flow velocimetry *Meas. Sci. Technol.* **21** 129–247
[36] Buchmann N A, Willert C E and Soria J 2012 Pulsed, high-power LED illumination for tomographic particle image velocimetry *Exp. Fluids* **53** 1545–60
[37] Maheo P 1998 Free-surface turbulent shear flows *PhD thesis* CalTech, Pasadena, CA
[38] Novara M and Scarano F Lagrangian acceleration evaluation for tomographic PIV: a particle-tracking based approach *16th Int Symp on Applications of Laser Techniques to Fluid Mechanics (Lisbon, Portugal, 9–12 July 2012)*
[39] Novara M and Scarano F 2013 A particle-tracking approach for accurate material derivative measurements with tomographic PIV *Exp. Fluids* **54** 1584
[40] Troolin D R and Longmire E K 2010 Volumetric velocity measurements of vortex rings from inclined exits. *Exp. Fluids* **48** 409–20
[41] Elsinga G E 2008 Tomographic particle image velocimetry and its application to turbulent boundary layers *PhD Thesis* TU Delft
[42] Elsinga G E, Wieneke B, Scarano F and Schröder A 2008 Tomographic 3D-PIV and applications *Topics Appl. Phys.* **112** 103–25
[43] Willert C E and Gharib M 1991 Digital particle image velocimetry *Exp. Fluids* **10** 181–93
[44] Hain R, Kähler C J and Tropea C 2007 Comparison of CCD, CMOS, and intensified cameras *Exp. Fluids* **42** 403–11

Chapter 3

Particle image identification

Particle tracking velocimetry (PTV) tracks individual particles to establish a velocity field. Its successful implementation therefore depends on its ability to accurately identify particles in a pair (or series) of images, and to properly match them. In this process, Feng *et al* [1] showed that there are two types of errors that affect PTV. First, particle accelerations due to either change in particle direction or speed can affect the accuracy of particle velocity measurements. Second, errors in particle locations can affect particle velocity measurements. In their study, they showed that the total velocity error is dominated by the particle position uncertainty for small particle displacements (small Δt), and for large particle displacements (large Δt), the total velocity error is dominated by particle accelerations. As most flows are imaged with small Δt between laser pulses, particle location uncertainties dominate the velocity error and, in this regard, this chapter focuses on particle image identification.

The first step in processing PTV images is identification of particle images for 2D and most 3D experiments. Some 3D techniques, such as tomographic PTV (section 4.2.2), synthetic aperture PTV (section 4.2.3), and plenoptic imaging (section 4.2.4), perform the particle identification step as part of the volumetric reconstruction step, which will be discussed in their respective sections.

There are two steps in particle identification: first an algorithm must determine which group of pixels constitutes an image of a particle; second the location of this particle image must be accurately determined. Accuracy in this step is of the utmost importance, as random errors in image position estimates propagate to both spatial position and velocity estimates. Accurate position estimates are also integral for reliable determination of depth coordinates using a multi-camera set-up (section 4.2).

The ideal raw frame from the CCD would contain bright particle images on a black background. Typically, however, noise exists in the images due to fluctuations in illumination intensity, video camera noise, and particle shadows. Section 3.1 discusses particle identification of these non-overlapped particles, which is

doi:10.1088/978-0-7503-2203-4ch3

applicable to lowly seeded flows. Additionally, some particle images overlap, causing ambiguity as to the number and location of present particles. A number of creative techniques have therefore been developed for extracting particle information and are discussed in section 3.2, where their development, implementation, and strengths are also described. This is followed by a comparison of published results in section 3.3.

3.1 Non-overlapped particles

3.1.1 Threshold binarization

A typical approach for identifying particle images is the use of segmentation. This is done using threshold binarization, which sets a pixel intensity to 1 if the gray level exceeds a set threshold and 0 otherwise [2]. Coherent groups of white pixels are then labeled as separate particle images. While this works for ideal cases, often illumination is inconsistent and bright background shapes can exceed the global threshold (figure 3.1a).

Thus, improvements were made using local threshold binarization [3] and dynamic threshold binarization [4]. Local threshold binarization determines a threshold based on the local average and variance of pixel intensity. Dynamic threshold binarization, on the other hand, uses a global base threshold. Adjacent pixels that exceed the threshold are then labeled together as an image and compared to the threshold value. If the contrast does not exceed a preset threshold, then the procedure for that image is finished; otherwise, one level of brightness is subtracted from each pixel in the area. This is repeated until all images have a contrast within the preset threshold (figure 3.1b).

3.1.2 Centroid estimation

Once identified in the CCD frame, an estimate of the particle's location must be found for tracking. This can be calculated as the area centroid of the pixels, defined as the ratio of the first-order moment to the zeroth-order moment, that make up the particle image. The binarized image contains separate segments of connected pixels. Each segment is considered a separate particle image, as shown below. Simple threshold binarization algorithms typically will calculate the mean x coordinate and

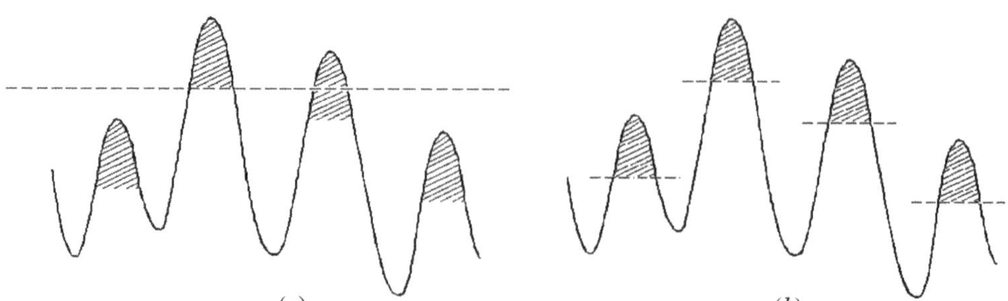

Figure 3.1. (a) Single threshold binarization and (b) dynamic threshold binarization [4].

mean y coordinate of each particle image to determine its location. The centroid calculations can also be weighted by pixel gray levels (figure 3.2).

The area centroid estimate is a non-parametric estimator of the particle image location. The estimate is unbiased when particle images are symmetrical. Systematic errors can be eliminated if the image's maximum spatial frequency illuminating the image sensor is less than the image sensor's sampling frequency, which can be obtained by setting the lens aperture so that its modulation transfer function cut-off frequency resulting from diffraction is smaller than the sensor's sampling frequency. Asymmetrical particle images introduce a systematic error in centroid estimation, and undersampling introduces a centroid estimation bias toward integer pixel values [6, 7].

3.1.3 Gaussian estimation

Once particle image blobs have been separated from the image background, particle positions in the pixel domain need to be determined in order to accurately measure the flow velocities for PTV. Although the particle images are pixelated, techniques for determining particle positions at a sub-pixel accuracy can be used. Initial studies using non-overlapped particle images investigated the parabolic, three-point center-of-mass, five-point center-of-mass, and three-point Gaussian estimators and identified the Gaussian estimator as the best estimator for particle positions [8]. In a later study, the three-point Gaussian (equations (3.1)) and a least-square Gaussian estimator (equation (3.2)) were studied, concluding that both results were comparable, but recommending the three-point Gaussian estimator as it was 100 times faster [9]. In both of these equations, G is the array of grayscale pixel intensities, (x_0, y_0) are the coordinates of the brightest pixel of a particle image, I_0 is the intensity, σ is the standard deviation of the Gaussian spread, and (x_c, y_c) are the real center positions. These algorithms are based on the assumption that particle images are Gaussian and circular or elliptical and axially oriented, which means the semi-major axis is parallel to the x or y direction. This is maintained for spherical particles with short exposure times compared to particle velocity. In situations with elliptical particle images that are not axially oriented, however, pixel-locking errors occur:

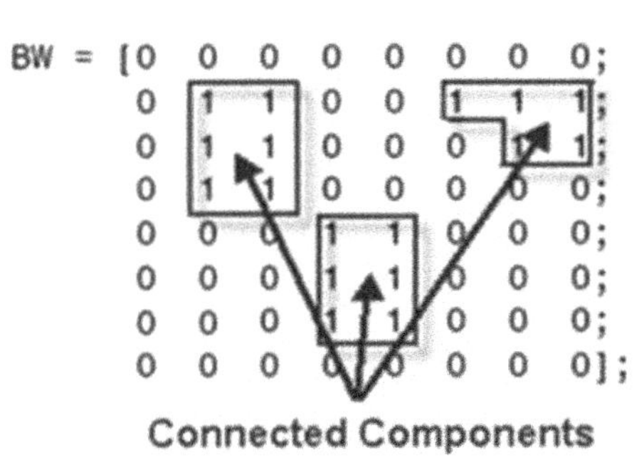

Figure 3.2. Binary matrix showing three separate particle images. Reprinted from [5] with permission of Springer.

$$x_c = x_0 + \frac{\ln G(x_0 - 1, y_0) - \ln G(x_0 + 1, y_0)}{2(\ln G(x_0 - 1, y_0) + \ln G(x_0 + 1, y_0) - 2\ln G(x_0, y_0))} \tag{3.1}$$

$$y_c = y_0 + \frac{\ln G(x_0, y_0 - 1) - \ln G(x_0, y_0 + 1)}{2(\ln G(x_0, y_0 - 1) + \ln G(x_0, y_0 + 1) - 2\ln G(x_0, y_0))} \tag{3.2}$$

$$\chi^2(x_c, y_c, I_0, \sigma) = \sum\nolimits_{i,j=-1}^{1} \left(G(x_i, y_i) - I_0 e^{-\frac{(x_i - x_c)^2 + (y_i - y_c)^2}{2\sigma^2}} \right)^2. \tag{3.3}$$

An improved method presented by Nobach and Honkanen [10] allows for an elliptical particle to be oriented in any direction so long as it maintains a Gaussian shape. The regression is compared visually to a standard three-point estimator in figure 3.3. The two-dimensional fit uses a least-square fit of a Gaussian as outlined in equation (3.2). The elliptical Gaussian intensity function can be reorganized, and the coefficients can be solved for a least-squares fit, yielding a maximum that would identify the particle position. This algorithm can effectively locate rotated elliptical images, although it cannot handle a background gray value. This means that a pre-processing is required to remove background gray offset. The performance of the two-dimensional Gaussian estimator with and without a low-pass filter is compared in figure 3.4. The simulated photon noise is based on the Poisson distribution of the number of photons, where 10 000 photons is the maximum intensity of a particle image.

Both methods have significant systematic deviations for particle diameters below 2 pixels, as the particle images in this range do not match the Gaussian model used. For particle image diameters greater than 3 pixels, and no photon noise, the two algorithms perform comparably. With the introduction of photon noise, bias error increases with image diameter, and the 2D estimator has less RMS bias than the 2 × 3 interpolation. For particle image diameters between 2 and 3 pixels, the nine-point regression has substantially higher bias than the 2 × 3 interpolation. The use of a Gaussian low-pass filter improved the accuracy of the nine-point regression for

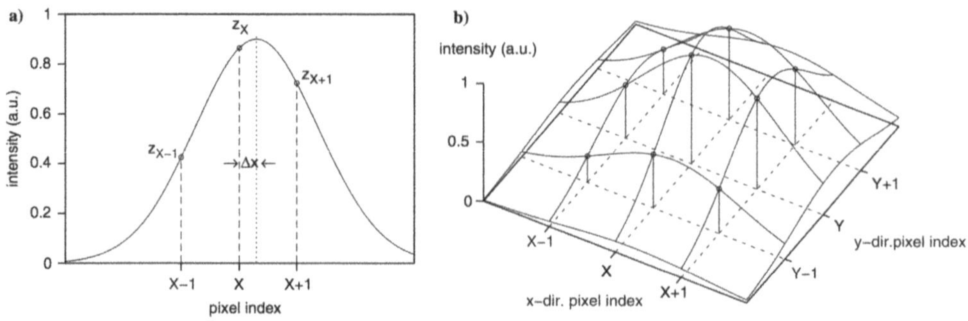

Figure 3.3. (a) One-dimensional three-point Gaussian estimator and (b) two-dimensional Gaussian regression. Reprinted from [10] with permission of Springer.

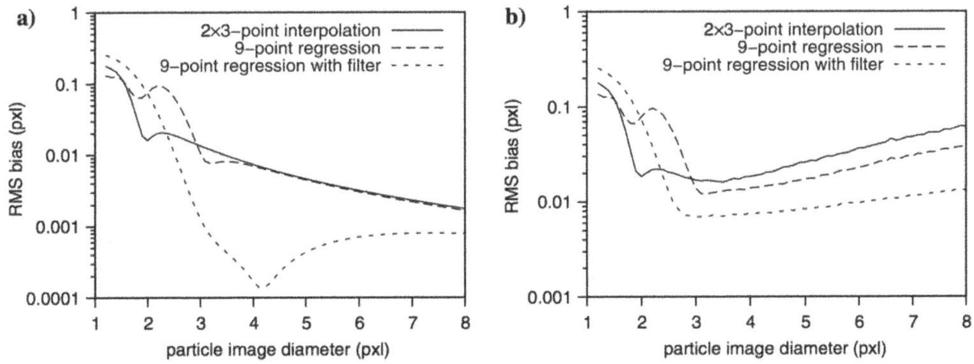

Figure 3.4. RMS error plotted for different particle position estimates (a) without noise and (b) with simulated photon noise. Reprinted from [10] with permission of Springer.

particle image diameters greater than 2 pixels. The low-pass filter reduces the noise and makes the particle image better resemble the assumed Gaussian shape.

3.2 Overlapped particles

In the studies described in the previous section, it was recognized that increasing the particle density resulted in increasing overlapped particles that were incorrectly identified as single particles. This resulted in an incorrect position estimates and thus erroneous results. While prevalent in highly seeded 2D applications, particle overlap is most common in 3D applications. According to Maas *et al* [11], for a 512 × 512 pixel image with 2000 particles (0.0076 particles per pixel or N_{ppp}) with each covering 10 pixels on average (or the average particle diameter is 3.6 pixels), the number of overlapped particles increases linearly with the particle image size and with the square of the number particles per image. For this case, 300 cases of overlapping particles were found. Consequently, many early PTV studies worked with sparsely seeded flows in order to avoid this limitation.

3.2.1 Threshold binarization

Many techniques have been proposed to identify overlapped particles. Maas *et al* [2] developed a modified anisotropic thresholding operator that looked for gray value discontinuities within segmented blobs, and if a discontinuity exceeded a given threshold, the operator split the segmented blobs. While this method was capable of finding multiple particle centroids within an image, the assumption that gray level increases continuously up to local peaks sometimes seemed too rigorous when one considers the sensitivity fluctuation of CCD sensors [4].

Mikheev and Zubtsov [12] modified the dynamic threshold segmentation proposed by Ohmi and Li [4] to detect the inflection point between two neighboring intensity peaks. This allowed them to identify multiple particle positions within segmented blobs.

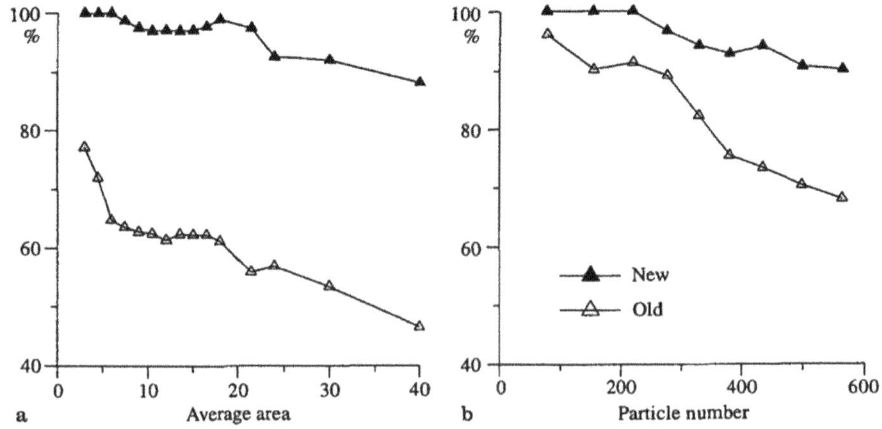

Figure 3.5. Correct barycenter recognition percent versus (a) average number of pixels per particle and (b) number of particles in a 512 × 512 pixel image. Reprinted from [13] with permission of Springer.

3.2.2 Neural network particle identification

Carosone *et al* [13] proposed a neural network for identification of overlapped particle images. Input sensors contain information about each particle image regarding its shape. The inputs interact with a set of neurons with weights for each input, outputting an activation value for each neuron. The neural network weights were taught with a learning data set composed of example images that are properly labeled as an overlapped or a single particle. The neuron with the minimum activation output is then selected and used to label the image. Half of the neurons classify the image as an overlap, while the other half classify the image as a single particle. The authors first defined a circularity measure, *crf*, as 4π times the particle's area divided by the particle's squared perimeter, and a convexity measure, *cnv*, as the ratio of the perimeter of the smallest possible convex shape that encloses the image and the actual shape perimeter, where both of these measures are equal to one for a circle. These were two inputs among many other defined parameters, and based on trials using synthetic data, the combination of these two inputs gave the highest percent of average recognition of particle images. Once particle images were labeled as overlapped, a boundary erosion algorithm was used to separate the individual particles. Once separated, the center of area of each particle image was calculated using the method outlined in the previous section.

Results for the neural network were compared to those of an algorithm that does not separate overlapped particles. The correct barycenter recognition percent was compared for these two algorithms based on synthetic images. Figure 3.5(a) varies the average number of pixels per particle image with a constant average displacement of 7 pixels and 600 particles (the largest particle concentration studied was 600 particles over a 512 [2] pixel array, or 0.0023 N_{ppp}). Figure 3.5(b) holds the average particle area constant at 12 pixels per particle image (or 4 pixel particle diameter) and varies the particle concentration. The neural recognizer outperforms the previous identification algorithm in all cases, but is not recommended for images with small number of overlaps or average particle areas below 7 pixels (or 3 pixel particle image diameter). Particle images that consist of only a small number of pixels will always be identified as overlapped according to the circularity and convexity measures used. Thus, the algorithm performs boundary erosion on every image and substantially increases the computational cost. The increase in performance is minimal when the number of overlapped particles is small, therefore the algorithm is not recommended due to the computational cost.

3.2.3 Particle mask correlation

The particle mask correlation algorithm by Takehara and Etoh [14] was introduced as an alternative to threshold binarization. The technique assumes the shape of particle images and searches for areas where the pixel intensity correlates with the assumed shape. Takehara and Etoh suggested that a Gaussian distribution, given in equation (3.3), best describes particle images, where a is the peak intensity, x_0 and y_0 give the image centroid coordinates, and σ gives a measure of the size. Equation (3.4) gives the normalized correlation coefficient calculated for an $m \times n$ pixel window with pixel intensity, I, mean pixel intensity, $\hat{I}$, mask intensity, I_m, and mean mask intensity, $\hat{I}_m$. Normalized correlation coefficients are then binarized, with an experimentally determined optimal threshold of 0.7. The area is extracted from this binarized image, whereafter the image's center of gravity and equivalent radius are then calculated:

$$I(x, y) = a^* e^{-((x-x_0)^2+(y-y_0)^2)/2\sigma^2} \tag{3.4}$$

$$r(x_0, y_0) = \frac{\sum_{x_0-m/2}^{x_0+m/2}\sum_{y_0-n/2}^{y_0+n/2}(I(i, j) - \hat{I})(I_m(i, j) - \hat{I}_m)}{\sqrt{\sum_{x_0-m/2}^{x_0+m/2}\sum_{y_0-n/2}^{y_0+n/2}(I(i, j) - \hat{I})^2}\sqrt{\sum_{x_0-m/2}^{x_0+m/2}\sum_{y_0-n/2}^{y_0+n/2}(I_m(i, j) - \hat{I}_m)^2}}. \tag{3.5}$$

It was determined that the particle mask method is capable of eliminating linear elements, such as boundary edges or immersed bodies, within the camera frame. Additionally, for a fixed σ value equal to 3.0 pixels, particle images between 0.3σ and 8σ were identifiable. For overlapping particles of similar size and brightness, the critical distance, L, for proper identification was determined to be $L = 2\sigma$. This critical distance increases as the difference between the particles' peak brightness increase, up to a maximum of $L = 4\sigma$ for a brightness ratio of 1/20.

An improvement upon the concept of the particle mask correlation method, called the cascade correlation method, was introduced by Angarita-Jaimes *et al* [15]. The technique initially performs the same cross-correlation shown in equation (3.4), labeling this as R_1. Then the correlation is repeated for R_2 by replacing I with R_1 in equation (3.4) and reducing the radius of the mask intensity distribution. Peak pixel intensities in the R_2 coefficient distribution are then found to determine the number and location of each particle. This technique introduces errors in position determination, as locations were calculated from the cross-correlation coefficient distribution rather than the pixel intensity distribution.

To improve upon the cascade correlation method, Lei *et al* [5] repeated the correlation process described above, each time with the particle mask radius reduced by one pixel. This iterative process was terminated when the mask's radius reached a user-defined minimum value. In addition, they decided to simply extract the identified number of peaks, N, from the cascade correlation method and use this information to perform a least-squares fit on the original pixel intensity distribution, shown in equation (3.5),

$$\chi^2 = \sum_{j=1,k=1}^{j=m,k=n} \left[I(j,k) - \sum_{i=1}^{N} I_o e^{\frac{x_i(j,k)^2}{2r_i^2}} \right]^2, \quad (3.6)$$

to determine the particle centroid locations. Figure 3.6 shows the flow diagram for the improved cascade correlation particle identification algorithm. Lei *et al* also improved the blob identification by using Otsu's method (1979) [16], which segments the foreground (blobs) and background image (noise) by finding the threshold to minimize the variance between the two classes of pixels.

3.2.4 Optical flow feature extraction

Shindler *et al* [17, 18] presented a new algorithm for feature extraction, which would be considered a particle, based in optical flow. The optical flow equation conserves the image intensity of a particle image according to

$$\begin{aligned} \frac{DI}{Dt} &= \frac{\partial I}{\partial t} + u\frac{\partial I}{\partial x} + v\frac{\partial I}{\partial y} = I_t + uI_x + vI_y \\ &= \frac{\partial I}{\partial t} + \nabla I^T \cdot \vec{U} = 0; \; \nabla I(\vec{x},t) = \begin{bmatrix} I_x \\ I_y \end{bmatrix}, \end{aligned} \quad (3.7)$$

where $\vec{U}$ is the unknown velocity vector. This problem is solved by reformulating it as a minimization in the least-squares sense, leading to the system of equations [19, 20]

$$\boldsymbol{G} \cdot \boldsymbol{U} + \boldsymbol{b} = \boldsymbol{0}; \quad \boldsymbol{U} = -\boldsymbol{G}^{-1} \cdot \boldsymbol{b}, \quad (3.8)$$

where

$$\boldsymbol{G} = \begin{bmatrix} \int_W I_x^2 dS & \int_W I_x I_y dS \\ \int_W I_y I_x dS & \int_W I_y^2 dS \end{bmatrix}; \quad \boldsymbol{b} = \begin{bmatrix} \int_W I_x I_t dS \\ \int_W I_y I_t dS \end{bmatrix}. \quad (3.9)$$

The solution requires that the eigenvalues, λ_1 and λ_2, of $\boldsymbol{G}$ be non-zero, and that they be real and positive. In this regard, a 'good feature to track', which would be considered a particle, would need to satisfy

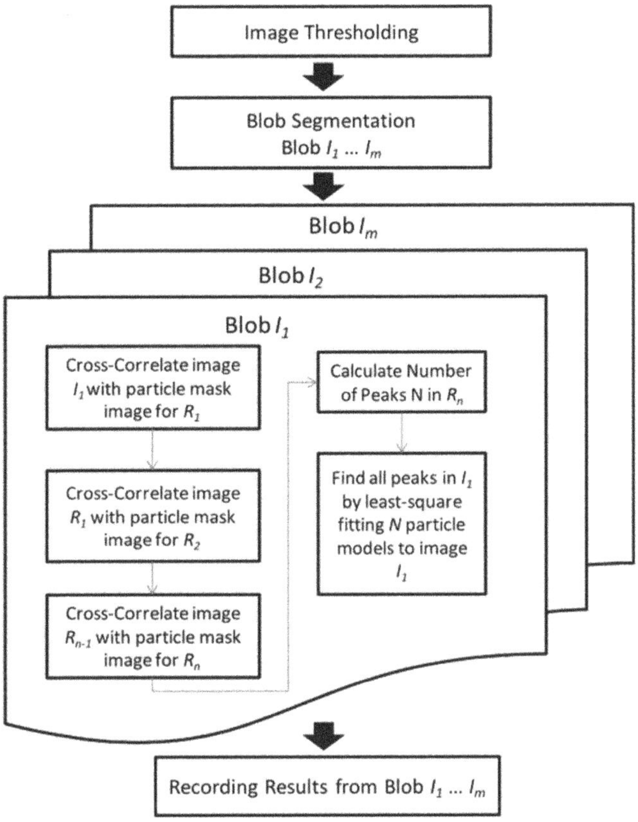

Figure 3.6. Flow chart for the modified cascade correlation particle identification algorithm. Reprinted from [5] with permission of Springer.

$$\min(\lambda_1, \lambda_2) > \lambda_t, \tag{3.10}$$

where

$$\lambda_t = C \cdot \max(\min(\lambda_1(x, y), \lambda_2(x, y))) \tag{3.11}$$

where C is a user-defined parameter not larger than unity. The value of C has been chosen between 0.005 and 0.01 for medium noisy images and should be increased for high levels of noise. A parameter characterizing the minimum distance between features is also given in their algorithms to allow for variations in seeding density, which then define the identified feature as one particle. The barycenter of this particle is then obtained by fitting a Gaussian fit, as given in equation (3.1), to their feature. In addition to identifying particles, the least-squares minimization of equation (3.7) gives an initial estimate of local velocity.

3.2.5 Linear model inversion

In order to identify particle locations within high particle density environments while minimizing errors, Cheminet *et al* [21] developed a method that reconstructs particles using a linear model inversion approach that accounts for both non-negativity and sparsity. First, they model the particle distribution within space and their multiview camera images as a linear relation,

$$\boldsymbol{Ax} = \boldsymbol{b}, \tag{3.12}$$

where $\boldsymbol{x}$ contains the particles' intensities within the imaged volume, $\boldsymbol{b}$ contains the multiview images' pixels, and $\boldsymbol{A}$ is the matrix that projects the 3D space into the 2D space of the cameras' images. This matrix also depends on the projection model. While tomographic PIV assumes that the pixel intensity is due to the integrated light along the projected line from particles in space to their images (see section 4.2.2 and figure 4.32), this work models their projection to be physics-based, such that particle positions and their intensities are identified within a few voxels, resulting in a sparse vector. Based on earlier work by Champagnat *et al* [22], the particles are modeled as point sources that result in an image intensity described as

$$I(\boldsymbol{x}) = \sum_{p=1}^{P} E_p h_{X_p}(\boldsymbol{x} - F(\boldsymbol{X}_p)), \tag{3.13}$$

where $\boldsymbol{P}$ are the particles, $\boldsymbol{E_p}$ are their intensities, X are their locations, x is the image coordinate, $\boldsymbol{F}$ is the function that projects particle positions from spatial coordinates to image coordinates, and $\boldsymbol{h}$ is the point spread function that models the particle images, which is approximated with a 2D Gaussian distribution. In practice, the image intensity is then discetized onto the camera pixel array. An inversion algorithm is then used to obtain the particle intensities. Theoretically, there are an infinite number of solutions. However, recognizing that the particle intensities are non-negative numbers and that the solution is expected to be sparse, Cheminet *et al* use the non-negative least-squares (NNLS) method, which makes use of the lsqnonneg MATLAB optimization routine, in order to obtain a solution.

3.3 Particle identification comparison

3.3.1 Non-overlapped particle identification comparison

Although at present PTV methodologies emphasize the development of algorithms that can detect overlapped particle positions accurately, it is important to evaluate the accuracy of non-overlapped particle position estimators. In doing so, developers will be able to compare the performance of their overlapped particle position estimators with the non-overlapped particle position estimators, as the latter will provide the best estimate which the former can be measured against.

Marxen *et al* [9] compared the Gaussian three-point and the Gaussian least-square particle center estimators and quantified the particle position uncertainties. In figure 3.7, they show the RMS errors as a function of particle diameter, for a

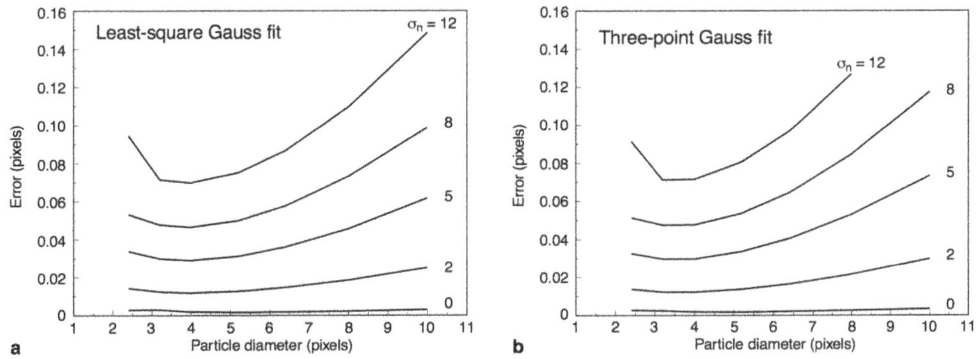

Figure 3.7. RMS error of the center location with respect to the true value of a function of particle diameter and the standard deviation of the Gaussian noise σ_n (maximum peak intensity = 200). Reprinted from [9] with permission of Springer.

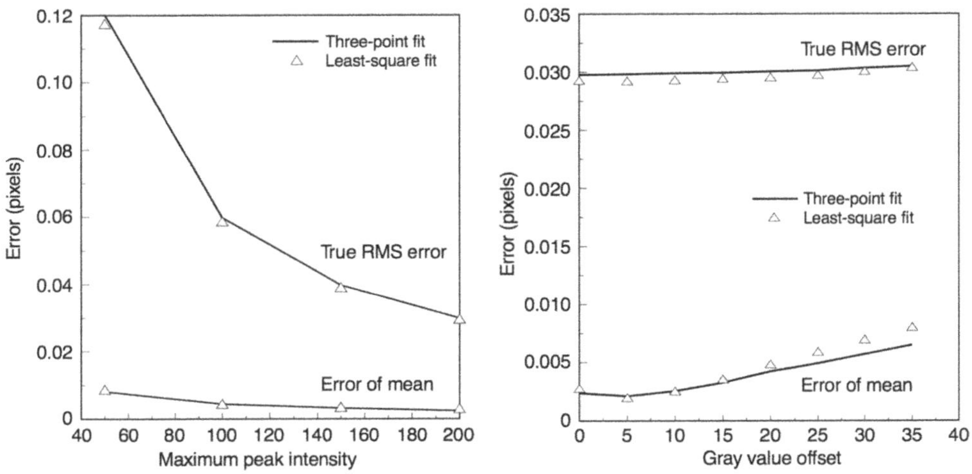

Figure 3.8. (left) RMS error and error of the mean value as a function of the maximum peak intensity ($\sigma_n = 5$, particle diameters = 4 pixels); (right) RMS error and the error of the mean value as a function of the maximum peak intensity ($\sigma_n = 5$, particle diameters = 4 pixels). Reprinted from [9] with permission of Springer.

range of Gaussian noise added to their images. They show that the errors are very similar, and that the effect of noise is significant.

In addition, they also studied the influence of different peak intensities and the offset of the Gaussian gray-value distribution from zero, using images with 4 pixel particle diameters and $\sigma_n = 5$. Their results, shown in figure 3.8, show that both the mean and RMS errors have a strong dependence on the peak intensities, while the errors are weakly dependent on the gray offset values. As discussed earlier in section 3.1.3, the authors recommended the use of the three-point Gaussian estimator as it was 100 times faster than the least-square Gaussian estimator.

3.3.2 Overlapped particle identification comparison

Ohmi and Li [4] used the PIV standard image 01 to compare three different particle position estimators, the results of which are shown in table 3.1. Here, the number of particles identified in the PIV standard image 01 for each method is listed [23]. In these results, it is shown that the dynamic threshold binarization algorithm was able to identify more particles, although uncertainties in particle positions were not reported and compared. In their modified dynamic threshold binarization algorithm, the mean error and standard deviation of the particle positions, although not measured directly, were estimated to be 0.05 and 0.1 pixels, respectively. It should also be noted that in a study performed by Cardwell *et al* [24], the authors found that the modified dynamic threshold binarization method did not perform robustly in identifying overlapped particles as compared to the particle mask method.

The improvements of the cascade correlation method over the particle mask correlation method is shown in figure 3.9, where a reduction of ~70% can be seen at maximum overlap. The improvements to the cascade correlation method made by Lei *et al* [5] are shown in figures 3.10 and 3.11. Here, figure 3.10 shows the effect of separation distances of two identical particles with a particle diameter of 20 pixels, where the particle diameter profile is Gaussian and is defined as four times the standard deviation of its Gaussian profile. Figure 3.10(a) shows the results for Poisson noise, where in comparison to the cascade correlation method, their algorithm results in particle location errors of 0.06 and 0.16 pixels for SNRs of 10 and 25, respectively, for maximum overlapped particles. Similarly, figure 3.10(b) shows that their algorithm for data with Gaussian noise, also for maximum overlapped particles, results in errors of 0.07 and 0.25 pixels.

The effect of particle size and Gaussian noise is shown in figure 3.11. Here, particle diameters were varied from 4 to 20 pixels. For a Gaussian noise level of 2.5%, figure 3.11(a) shows that the algorithm requires the particle diameters to be greater than 10 pixels in order to resolve overlaps up to 50%, since particles with smaller diameters do not provide enough spatial distribution and hence resolution to be able to accurately resolve overlaps of up to 50%. For a Gaussian noise level of

Table 3.1. Comparison of identification algorithms for the PIV standard image 01 [4].

Particle identification algorithm	Number of particles identified/total particles	Identified N_{ppp} /total N_{ppp}
Single threshold binarization	1160/4000	0.018/0.06
Dynamic threshold binarization	1269/4000	0.019/0.06
Particle mask correlation	1134/4000	0.017/0.06
Optical flow feature extraction[a]	1330/4000	0.020/0.06

[a] The authors report on 51 images from the VSJ 301-series. The values here correspond to the averages of images 1 and 2 in order to be consistent with the other comparisons. The authors report an average identified particle number ~1330.

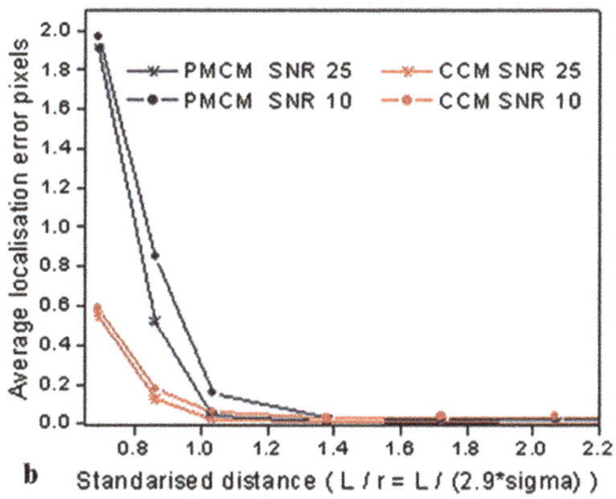

Figure 3.9. Comparison of the cascade correlation method with the particle mask correlation method [15].

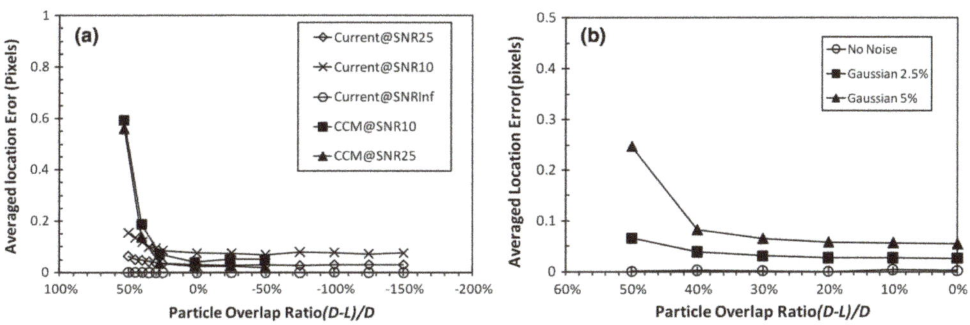

Figure 3.10. Averaged particle location error versus particle overlap ratio for (a) Poisson noise levels at infinity, 25, and 10, and (b) Gaussian noise levels at 2.5% and 5%. Reprinted from [5] with permission of Springer.

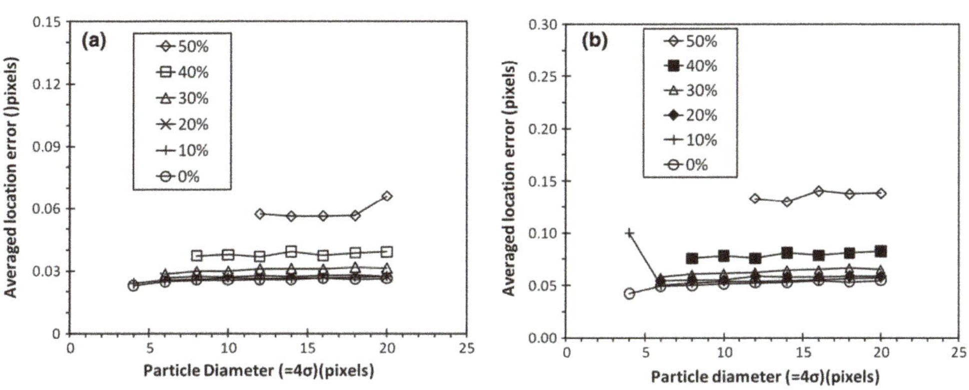

Figure 3.11. Averaged particle location error versus particle diameter for different overlap ratios: (a) 2.5% Gaussian noise and (b) 5% Gaussian noise. Reprinted from [5] with permission of Springer.

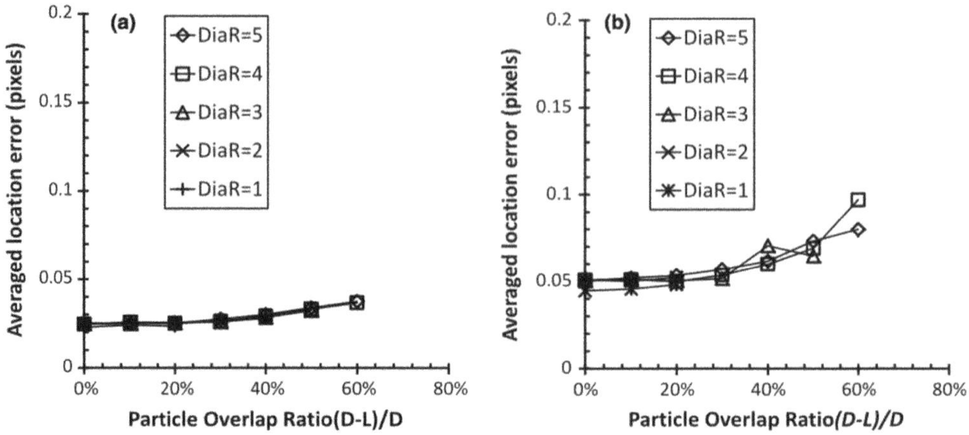

Figure 3.12. Averaged particle location error with various particle diameter ratios and noise levels: (a) 2.5% Gaussian noise versus particle overlaps and (b) 5% Gaussian noise versus particle overlap ratio. Reprinted from [5] with permission of Springer.

5%, figure 3.11(b) shows similar trends, however, the errors roughly double due to the doubled noise level. What is also seen is that larger particles result in higher particle location errors. For example, 6 pixel diameter particles have location errors of 0.03 and 0.06 for Gaussian noise levels of 2.5% and 5%, respectively, while 12 pixel diameter particles have location errors of 0.06 and 0.13 for the same Gaussian noise levels. For the 6 pixel diameter particles, these are 95% and 90% smaller than the Cascade correlation method for a Gaussian noise level of 2.5%, and 90% and 78% smaller than the Cascade correlation method for a Gaussian noise level of 5%.

Recognizing that in practice, seeding particles will vary in size, Lei *et al* also carried out similar studies with overlaps of particles with varying sizes at different Gaussian noise levels (see figure 3.12). For this study, their particle diameter was defined as the average of the two overlapped particles, where the smaller particle was set to 4 pixels in diameter. The diameter ratio was varied from 1 to 5. For 2.5% Gaussian noise, figure 3.12(a) shows that the curves overlap; however, only for diameter ratios of 4 and 5, the curves extend out to resolvable overlaps of up to 60%, with errors under 0.05 pixels. Figure 3.12(b) shows similar trends for 5% Gaussian noise, but with increased noise that is below 0.1 pixels.

It is important to point out that Lei *et al* [5] recognized that the VSJ overlapped particle images [23] used only the intensities of the brightest particle when forming the overlapped particle images. They showed that this was not the correct representation of overlapped particles, rather the correct representation of overlapped particles should sum all the intensities for each pixel from each of the overlapped particles, as is shown in figure 3.13. Here, it can be clearly seen that the experimental data and the superposition of the particles are much closer in profile than the method used to calculate the VSJ images. As a consequence, the VSJ images do not capture overlapped particle occlusions well and therefore the use of these images will artificially result in lower particle position uncertainty estimates.

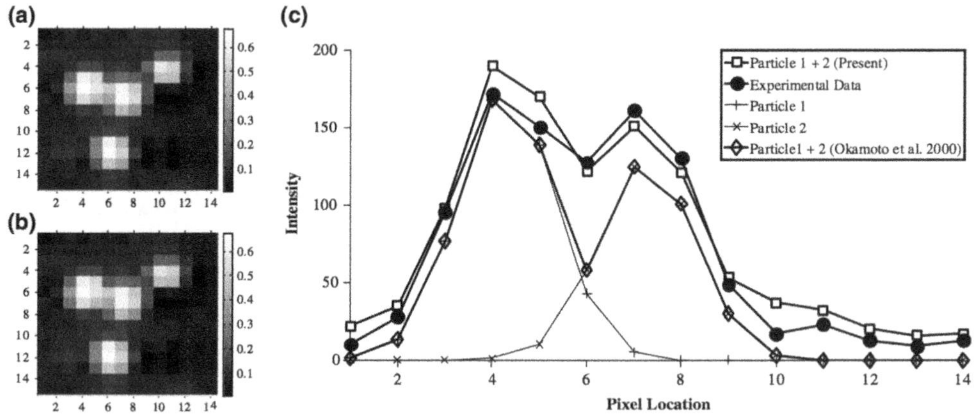

Figure 3.13. Comparison of particle image reconstruction performance based on different assumptions. (a) Original image, (b) reconstructed image by current work, and (c) comparison of profiles of the sixth row. Reprinted from [5] with permission of Springer.

TSI, Inc[1] also implemented the overlapping particle scheme developed by Lei *et al*, and studied it for use with $N_{\text{ppp}} \approx 0.1$. Rather than using Otsu's method to identify blobs, they use an erosion/dilation thresholding method, as proposed by Cardwell *et al* [24]. Local peaks within blobs are identified in the erosion step, and adjacent pixels assigned to each local maximum are identified in the dilation step, which with the particle peak were identified as 'support sets'. Rather than identifying each local peak within each support set as a particle location, TSI implemented Lei *et al*'s method on each support set to identify multiple peaks within each support set. Calling this approach dense particle identification and reconstruction (DPIR) and using synthetic particle images with 3 pixel diameters, DPIR showed that their results matched IPR's approach up to $N_{\text{ppp}} = 0.05$, but outperformed IPR by 39% for larger seeding density values (see figure 3.14). Also, DPIR's fraction of generated ghost particles was 80% lower than that generated with IPR (see figure 3.15). Finally, DPIR showed 48% lower reconstruction error (see figure 3.16) [25].

Champagnat *et al* [22] tested their linear model inversion method for particle reconstruction by using generated synthetic images to simulate randomly dispersed particles with 1 μm diameters within an illuminated volume. The laser sheet intensity was modeled to vary as

$$I_0 = 255\,\exp(-z^2/\sigma_{\text{laser}}^2), \tag{3.14}$$

where $2\sigma_{\text{laser}}$ is the laser sheet thickness, which was set to 1 mm. In these studies, the largest number of particles per pixel, N_{ppp}, was 0.11, and the particle diameter varied from 0.8 pixels to 4.8 pixels. To assess the algorithm's performance, they compared its results with those obtained by processing these images with the dynamic threshold

[1] https://www.tsi.com/products/fluid-mechanics-systems/volumetric-particle-image-velocimetry-systems/v3v-flex-volumetric-piv-systems/

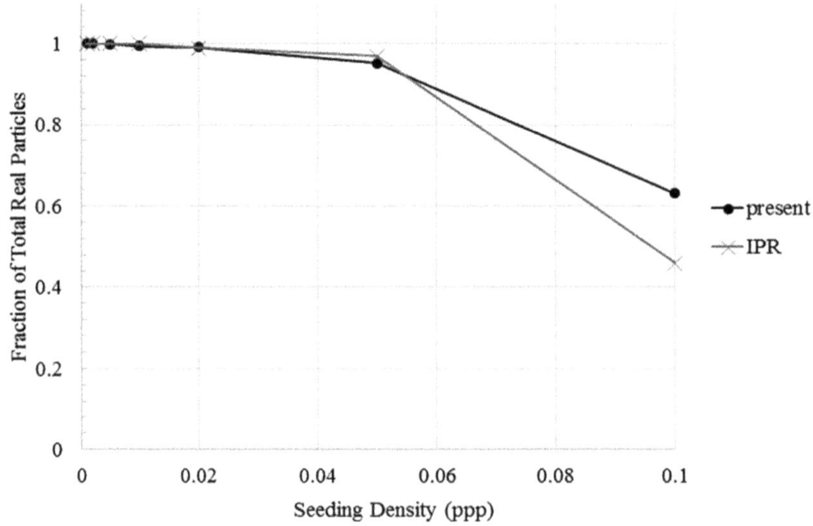

Figure 3.14. Fraction of total real particles identified versus seeding density. Reprinted from [25].

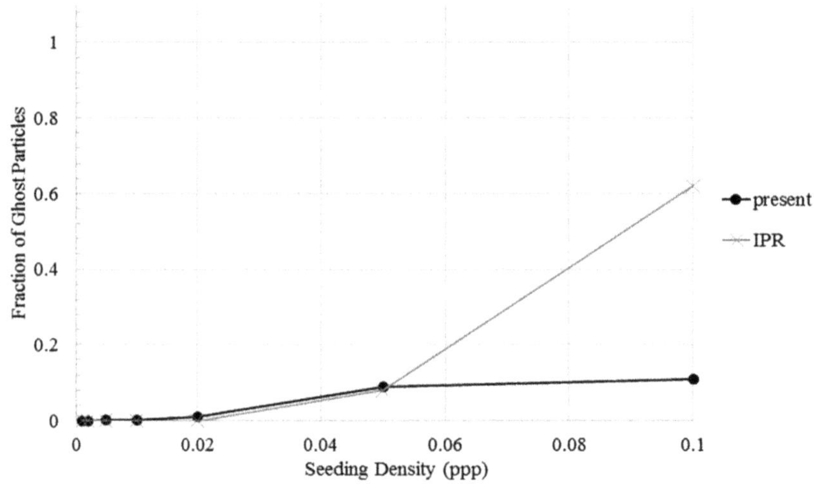

Figure 3.15. Fraction of ghost particles versus seeding density. Reprinted from [25].

binarization (DTB) [4], and to the modified cascade correlation method (CCM) [5]. The given acronym for their algorithm is PIR_{NNLS}, which stands for 'particle image reconstruction-non-negative least-squares'. To access the performance of these methods, two metrics, previously defined by Ruhnau *et al* [26] were used, the yield and the reliability. These are defined as

$$\text{Yield} = E_Y = \frac{\#\text{TP}}{\#\text{TP} + \#\text{FN}} \quad \text{Reliability} = E_R = \frac{\#\text{TP}}{\#\text{TP} + \#\text{FP}}, \tag{3.15}$$

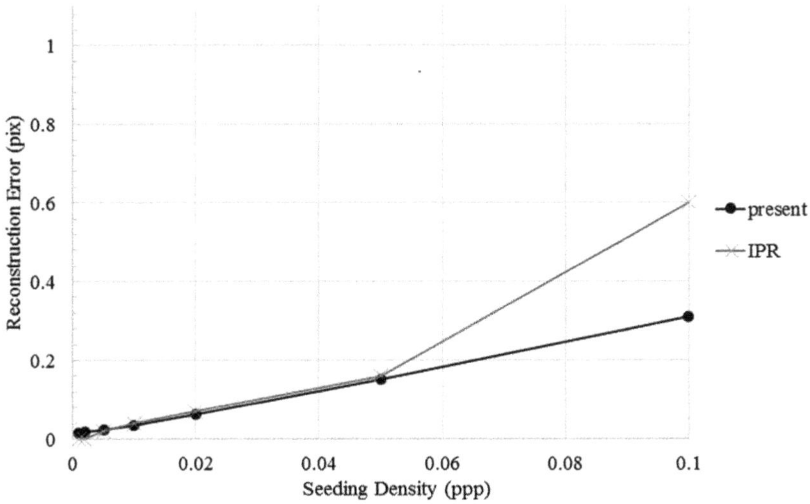

Figure 3.16. Reconstruction error in pixels versus seeding density. Reprinted from [25].

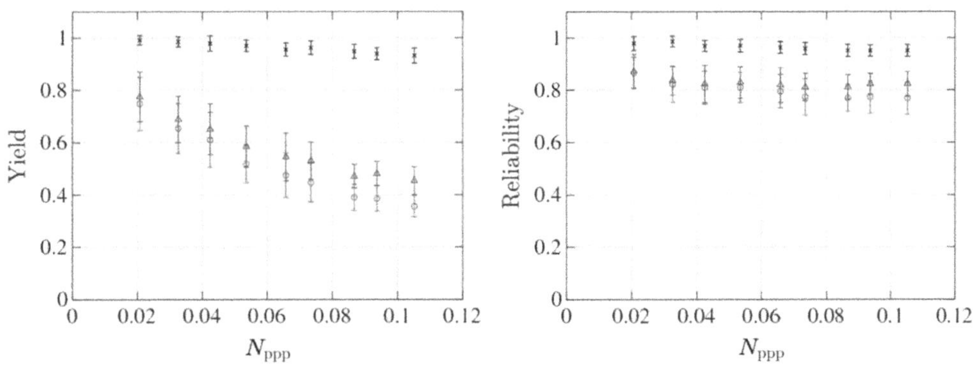

Figure 3.17. Comparison between the conventional DTB (blue Δ), CCM (red ○) and the proposed PIR_{NNLS} (black ×) in terms of detection efficiency as a function of the particle density, N_{ppp}. Without any camera noise. Left: yield; right: reliability. Reprinted from [21].

where TP is a true positive detection that is in the neighborhood of a particle and is the nearest detection to this particle, FP is a false positive detection that is not the nearest detection of a particle if is not in the neighborhood of any particle or inside the neighborhood of a particle, and FN is a false negative detection if there is no detection in its neighborhood. In this manner, the yield captures the fraction of true detections relative to the generated particles, while the reliability captures the fraction of true detections amongst all the detections.

Figure 3.17 shows mean results for a particle field of the yield and reliability for the three algorithms without any camera noise. On the left, the PIR_{NNLS} method clearly shows the highest yield, slightly dropping from almost 100% at $N_{ppp} = 0.02$ to ~95% at $N_{ppp} = 0.11$, while the other two methods show a decreasing yield with increasing N_{ppp}; at the largest N_{ppp}, the DTB and the CCM produce at best half the yield that the PIR_{NNLS} method delivers. While the reliability results are much closer, the PIR_{NNLS} method maintains values above ~98% for the range of N_{ppp} studied, while the DTB and CCM methods maintain values at ~80%. Also note that the standard deviation bars of both the yield and reliability plots for the PIR_{NNLS} method are about half of those shown for both of the DTB and the CCM methods.

Figure 3.18 shows mean results for a particle field of the mean localization errors in pixels for both the noiseless and noisy cases, where noise was generated by adding 5% Gaussian noise. In the noiseless case, it can be clearly seen that the PIR_{NNLS} method has the lowest mean noise, which is within 0.07 pixels for the range of tested N_{ppp}. The CCM method shows that mean noise increases from ~0.07 pixels at 0.02 N_{ppp} to ~0.18 pixels at 0.11 N_{ppp}. Finally, the DTB method has the largest mean noise, increasing from ~0.25 pixels at 0.02 N_{ppp} to ~0.3 pixels at 0.11 N_{ppp}. Note also that in the noiseless case, the standard deviation bars for the PIR_{NNLS} method are about half of those shown for the CCM and DTB methods. For the noisy case, it is seen that the PIR_{NNLS} and CCM methods produce comparable results. For the first two lowest N_{ppp} results, the CCM method shows ~0.03 pixel lower mean localization error, for $N_{\text{ppp}} \approx 0.044$, the mean errors are almost identical, and for larger N_{ppp} values, the mean localization errors are on average ~0.12 pixels larger for the CCM method compared to the PIR_{NNLS} method. The DTB method shows the largest mean localization error, ranging from ~0.25 pixels for the smallest N_{ppp} value to ~0.3 pixels for the largest N_{ppp} value. The standard deviation bars for all three cases are seen to be very comparable.

Figure 3.19 shows mean results for a particle field of the yield and reliability for the three algorithms with 5% added Gaussian noise. On the left, the PIR_{NNLS} method clearly shows the highest yield, ranging from ~96% at the lowest N_{ppp} value to ~84% at the largest N_{ppp} value. The CCM and DTB results are comparable, although their yields are noticeably lower than the PIR_{NNLS} method; these values range from ~74% at the lowest N_{ppp} value to ~37% at the largest N_{ppp} value. The

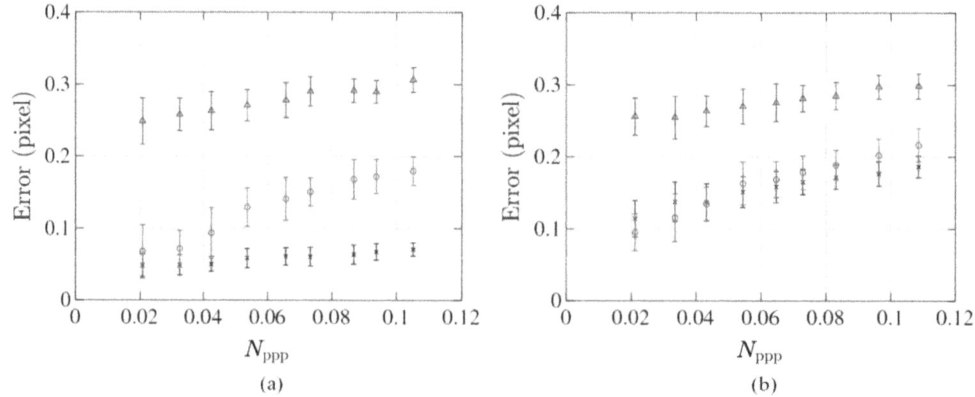

Figure 3.18. Comparison between the mean localization errors as obtained by PIR_{NNLS} (black ×), DTB (blue Δ), and CCM (red ○) as a function of the particle density, N_{ppp}. (a) Images without any noise. (b) Images with an added 5% Gaussian noise [21].

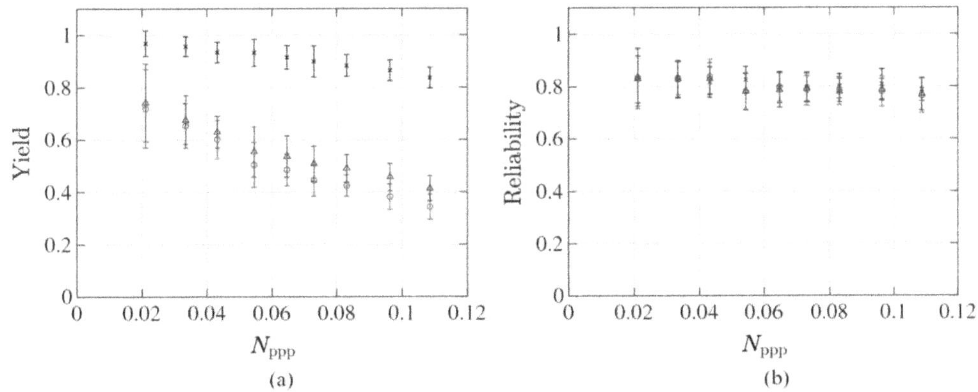

Figure 3.19. Quality of the detection as given by PIR_{NNLS} (×), DTB (Δ), and CCM (○) as a function of the particle density, N_{ppp}. With a 5% Gaussian noise. Left: yield; right: reliability [21].

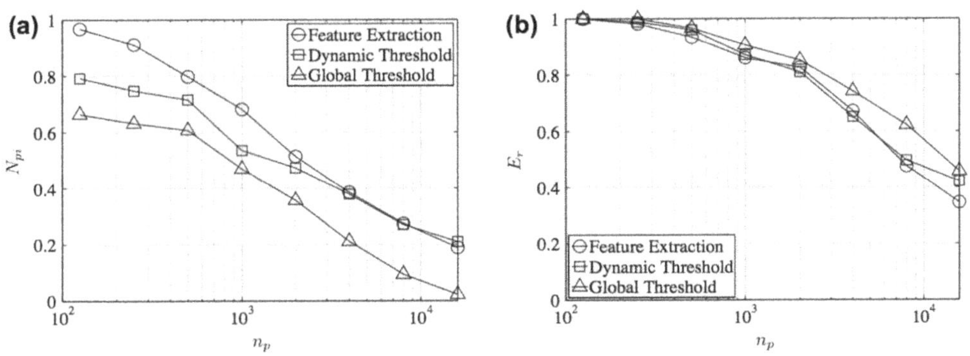

Figure 3.20. Tracer density influence on (a) the normalized number of identified particles, N_{pi}, and (b) the reliability, E_r [18].

reliability mean and standard deviation results for all three methods are very comparable, maintaining values at ~80% across the range of N_{ppp} values tested. It is important to note that for values of N_{ppp} ~5, the PIR_{NNLS} method easily doubles the vector yield, while maintaining consistent reliability values, thereby making this algorithm a promising method for accurate particle identification. The authors did also note that, although the reliability of the PIR_{NNLS} method is the same percentage as the CCM and DTB methods, since the PIR_{NNLS} method does produce about twice the yield, the percentage of false positive detections is larger and is due to the mathematical difficulty in performing the inversion algorithm.

The optical flow feature extraction (OFFE) technique is also compared with the dynamic threshold (DT) and global threshold (GT) methods [18]. Figure 3.20 shows the effect of particle density on the normalized particle density (identified particles per total particles, N_{pi}) and on the reliability, E_r, also defined as the number of correctly identified particles divided by the total number of identified particles. For these tests, the number of particles were varied from 125 to 16 000 (labeled n_p) within a 256 × 256 pixel image size, corresponding to a range of 0.002 to 0.244 N_{ppp}, with a

3 pixel mean particle diameter and a background mean and standard deviation Gaussian noise of 10 and 1, respectively. Figure 3.20(a) shows that the OFFE outperforms the GT by about 0.2 N_{pi} throughout the range of particle densities tested in identifying particles. At 0.002, .0038, 0.0076, 0.015, 0.03, and 0.06 N_{ppp}, the N_{pi} are about 0.975, 0.91, 0.8, 0.68, and 0.51, respectively. Beyond 0.06 N_{ppp}, the performances become identical. Figure 3.20(b) shows that of these identified particles, the reliability of OFFE is comparable to that of DT up to 0.12 N_{ppp}.

Figure 3.21 shows the effect of particle image radius on N_{pi} and E_r. For these tests, the mean particle image radius varied from 1 to 8, with 2000 particles per image, or 0.03 N_{ppp}, and a mean and standard deviation in the background Gaussian noise of 10 and 1, respectively. Figure 3.21(a) again shows that the OFFE outperforms the DT and GT, and that the OFFE's performance is mostly insensitive to the particle image radius. Of these identified particles, all methods show a reliability that is mostly similar up to a particle radius of 5 pixels (see figure 3.21(b)), after which the GT noticeably outperforms the DT and OFFE.

Figure 3.22 shows the effect of background mean noise on N_{pi} and E_r. For these tests, the mean particle image radius was set to 3, with 2000 particles per image, or 0.03 N_{ppp}. The mean background noise is labeled as F, and the standard deviation of the noise was set to 1. Figure 3.22(a) shows that all three algorithms' performance is mostly uniform across the noise span tested, where the OFFE shows the best performance. Figure 3.22(b) shows that even though the signal-to-noise decreases as the noise level goes up, that all three algorithms' performance degrades slowly, as the reliability gently decreases with increasing noise, where the OFFE outperforms the DT and the GT.

Figure 3.23 shows the effect of background standard deviation noise on N_{pi} and E_r. Similar to previous tests, the mean particle image radius was set to 3, with 2000 particles per image, or 0.03 N_{ppp}. The background standard deviation noise is labeled as σ_g. Also, the coefficient C in equation (3.10) was increased for the noisiest images in order to prevent false detections. Figure 3.23(a) shows that all three algorithms' performance is mostly uniform across the noise span tested, where the OFFE shows the best performance of the three methods tested. Figure 3.23(b) shows that for increasing standard deviation of the noise, the reliability of all the tested algorithms is minimally affected, and the GT outperforms OFFE and DT.

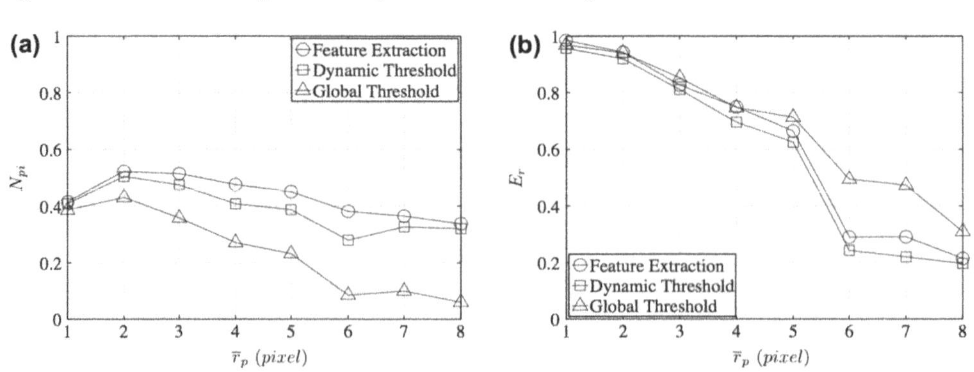

Figure 3.21. Particle radius influence on (a) the normalized number of identified particles, N_{pi}, and (b) the reliability, E_r [18].

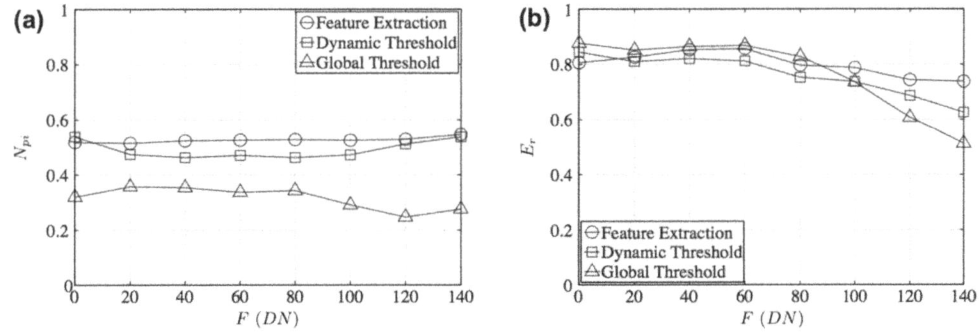

Figure 3.22. Background mean noise influence on (a) the normalized number of identified particles, N_{pi}, and (b) the reliability, E_r [18].

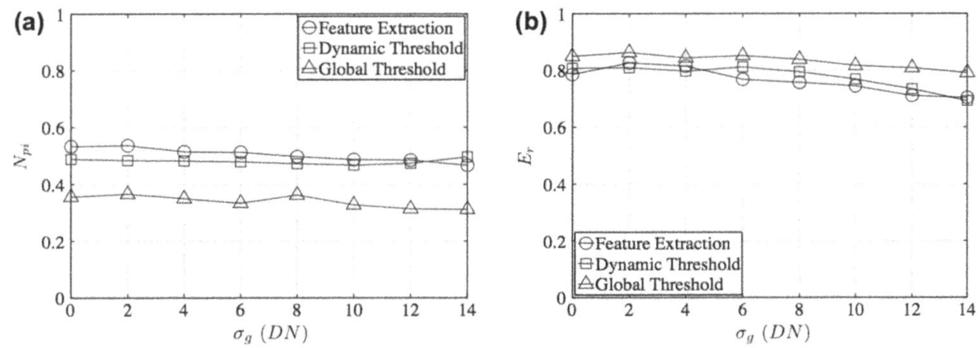

Figure 3.23. Background standard deviation noise influence on (a) the normalized number of identified particles, N_{pi}, and (b) the reliability, E_r [18].

References

[1] Feng Y, Goree J and Liu B 2011 Errors in particle tracking velocimetry with high-speed cameras *Rev. Sci. Instrum.* **82** 053707

[2] Maas H G, Gruen A and Papantoniou D 1993 Particle tracking velocimetry in three-dimensional flows Part 1. Photogrammetric determination of particle coordinates *Exp. Fluids* **15** 133–46

[3] Dezso-Weidinger G, Stitou A, van Beeck J and Riethmuller M L 2003 Measurement of the turbulent mass flux with PTV in a street canyon *J. Wind Eng. Ind. Aerodyn.* **91** 1117–31

[4] Ohmi K and Li H-Y 2000 Particle-tracking velocimetry with new algorithms *Meas. Sci. Tech.* **11** 603–16

[5] Lei Y-C, Tien W-H, Duncan J, Paul M, Ponchaut N, Mouton C, Dabiri D, Rösgen T and Hove J 2012 A vision-based hybrid particle tracking velocimetry (PTV) technique using a modified cascade-correlation peak-finding method *Exp. Fluids* **53** 1251–68

[6] Alexander B F and Ng K C 1991 Elimination of systematic error in subpixel accuracy centroid estimation *Opt. Eng.* **30** 1320–31

[7] Adrian R J and Westerweel J 2011 *Particle Image Velocimetry* (New York: Cambridge University Press)

[8] Cowen E A and Monosmith S G 1997 A hybrid digital particle tracking velocimetry technique *Exp. Fluids* **22** 199–211

[9] Marxen M, Sullivan P E, Loewen M R and Jahne B 2000 Comparison of Gaussian particle center estimators and the achievable measurement density for particle tracking velocimetry *Exp. Fluids* **29** 145–53

[10] Nobach H and Honkanen M 2005 Two-dimensional Gaussian regression for sub-pixel displacement estimation in particle image velocimetry or particle position estimation in particle tracking velocimetry *Exp. Fluids* **38** 511–5

[11] Maas H G, Gruen A and Papantoniou D 1993 Particle tracking velocimetry in three-dimensional flows Part 1. Photogrammetric determination of particle coordinates *Exp. Fluids* **15** 133–46

[12] Mikheev A V and Zubtsov V M 2008 Enhanced particle-tracking velocimetry (EPTV) with a combined two-component pair-matching algorithm *Meas. Sci. Technol.* **19** 085401

[13] Carosone F, Cenedese A and Querzoli G 1995 Recognition of partially overlapped particle images using the Kohonen neural network *Exp. Fluids* **19** 225–32

[14] Takehara K and Etoh T 1999 A study on particle identification in PTV—particle mask correlation method *J. Visualization* **1** 313–23

[15] Angarita-Jaimes N C, Roca M G, Towers C E, Read N D and Towers D P 2009 Algorithms for the automated analysis of cellular dynamics within living fungal colonies *Cytometry* A **75** 768–80

[16] Otsu N 1979 Threshold selection method from gray-level histograms *IEEE Trans. Syst. Man Cybern.* **9** 62–6

[17] Shindler L, Moroni M and Cenedese A 2010 Spatial–temporal improvements of a two-frame particle-tracking algorithm *Meas. Sci. Technol.* **21** 115401

[18] Shindler L, Moroni M and Cenedese A 2012 Using optical flow equation for particle detection and velocity prediction in particle tracking *Appl. Math. Comput.* **218** 8684–94

[19] Lucas B and Kanade T 1981 An iterative image registration technique with an application to stereo vision *Proc. Int. Joint Conf. on Artificial Intelligence* pp 674–9

[20] Tomasi C and Kanade T 1991 Shape and motion from image streams: a factorization method—Part 3. Detection and tracking of point features *Technical report* CMU-CS-91-132 Carnegie Mellon University

[21] Cheminet A, Krawczynski J-F and Druault P 2018 Particle image reconstruction for particle detection in particle tracking velocimetry *Meas. Sci. Technol.* **29** 125202

[22] Champagnat F, Cornic P, Cheminet A, Leclaire B, Besnerais G L and Plyer A 2014 Tomographic PIV: particles versus blobs *Meas. Sci. Technol.* **25** 084002

[23] Okamoto K, Nishio S, Kobayashi T and Saga T 1997 Standard images for particle imaging velocimetry *Proc. of the 2nd Int. Workshop on PIV '97- Fukui* pp 229–36

[24] Cardwell N D, Vlachos P P and Thole K A 2011 A multi-parametric particle-pairing algorithm for particle tracking in single and multiphase flows. *Meas. Sci. Tech.* **22** 105406

[25] Boomsma A and Troolin D 2018 Dense particle identification and reconstruction: DPIR *Presentation given at 19th Int. Symp. on Application of Laser and Imaging Techniques to Fluid Mechanics* (16–19 July)

[26] Ruhnau P, Guetter C, Putze T and Schnörr C 2005 A variational approach for particle tracking velocimetry *Meas. Sci. Tech.* **16** 1449–58

Dana Dabiri and Charles Pecora

Chapter 4

Identification of particles' spatial locations

While tomographic (section 4.2.2), synthetic aperture (section 4.2.3), plenoptic imaging (section 4.2.4), and holographic (section 4.2.5) methods directly identify particle spatial locations from reconstructed volumes, other PTV algorithms use particle positions within the image space to determine the corresponding positions in space. This then allows for tracking of particle tracers that will result in velocity vector fields in terms of real-world coordinates. The technique for identifying particle spatial locations varies between 2D and 3D methods. Section 4.1 therefore discusses 2D spatial localizations, while section 4.2 discusses 3D spatial localizations. Most of these transformations from the camera domain to the spatial domain require a calibration as well as methods for handling optical distortions, therefore calibration methods are discussed in section 4.3.

4.1 Spatial location in 2D

When only two-dimensional position and velocity are of interest, geometry can be used to determine the spatial coordinates from the camera plane. Assuming the pinhole camera model [1], equation (4.1) can be used to transform the image coordinates, X and Y, to the spatial coordinates, x and y. In this case, Z_0 is the distance from the image to the effective center of the lens and z_0 is the distance from the object to the effective center of the lens (this geometry is shown in figure 4.1):

$$\left(\frac{X}{Y}\right) = \frac{Z_0}{z_0}\left(\frac{x}{y}\right). \tag{4.1}$$

The pinhole model works for 2D imaging without any optical distortion or out-of-plane measurements. When out-of-plane motion exists, the following perspective error, given by Prasad [2], arises in the measurement of x displacement,

$$\varepsilon_x = \frac{\Delta z}{\Delta x}\sin\varphi_x, \tag{4.2}$$

doi:10.1088/978-0-7503-2203-4ch4

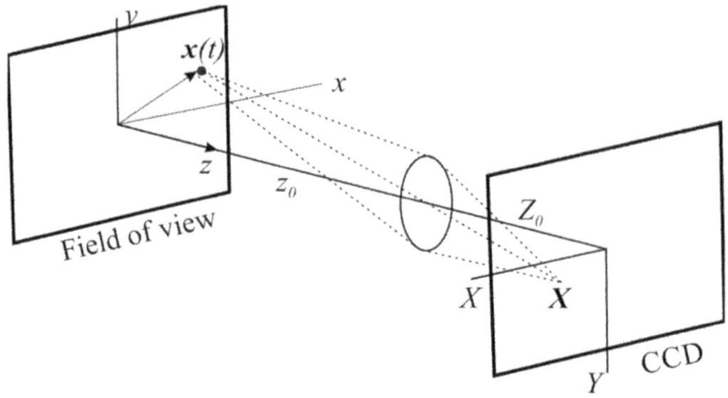

Figure 4.1. The geometry used in the pinhole model for determining 2D spatial coordinates of particles. Reproduced from [1], with permission from AIP Publishing.

where Δz is the out-of-plane displacement, Δx is the in-plane displacement, and φ is the viewing angle with respect to the optical axis. The uncertainties in 2D localization come only from errors in the image identification, calibration, and perspective. The calibration steps in section 4.3 can be used to address these uncertainties.

4.2 Spatial localization in 3D

The localization of particle coordinates in three dimensions can be a complex task. In order to extract three components of both position and velocity vectors, one of the methods outlined in this section is typically used. Rather than using only one camera, these 3D techniques incorporate multiple cameras and illuminate a volume rather a sheet. Due to this complexity, calibration steps are typically required in order to ensure proper localization.

4.2.1 Photogrammetric PTV

Photogrammetry is the science of making measurements from photographs. The underlying principle is that the distance between two points in an image can be determined if the scale is known. In the context of PTV, this scale depends on the depth of a particle image. This section outlines techniques that have been used to estimate depth and obtain 3D coordinates of imaged particles.

4.2.1.1 Multi-camera set-up

Most photogrammetric PTV techniques rely on multiple viewing angles in order to determine 3D coordinates. For extremely low particle densities, two cameras sufficiently provide all three spatial dimensions; however, a third or fourth camera can remove ghosts that arise when particle images overlap, as well as reduce uncertainties in the measurements. It should also be noted that stereoscopic flow imaging (two cameras) is most often used to obtain data on three-component velocities within a two-dimensional plane. Figure 4.2 shows an experimental set-up with two cameras for stereoscopic PTV and three cameras and volume illumination for 3D-PTV.

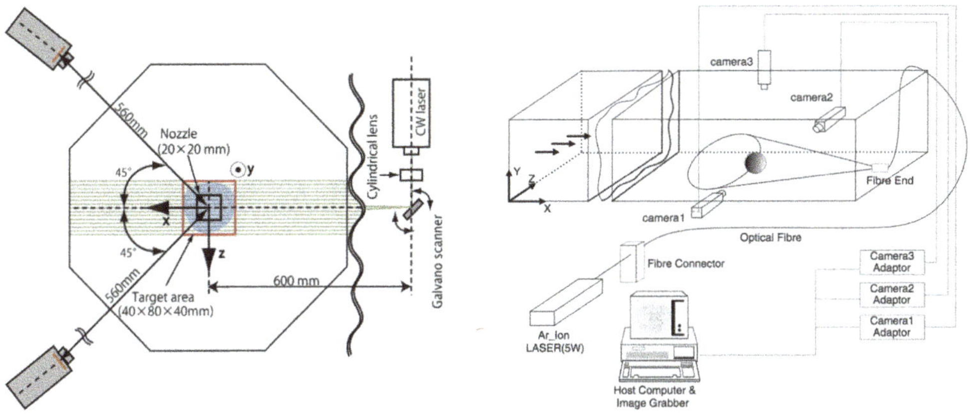

Figure 4.2. Set-up for stereoscopic PIV used to investigate a jet flow by Watanabe *et al* (left). Reprinted from [3], Copyright (2015), with permission from Elsevier. 3D-PTV set-up used to investigate the wake of a sphere by Doh *et al* (right) [4].

Table 4.1. Camera set-ups for 3D-PTV, their respective error factors, E, and sensitivity coefficients, B, for in-plane measurements. Δc is the horizontal distance between cameras and Δd is the vertical distance between cameras [5].

Set-up	Δc	Δd	B/M	E
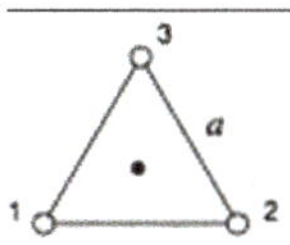	$-a, a/2, a/2$	$0, -a\sqrt{3}/2, a\sqrt{3}/2$	a	1
	$a, a, 2a$	0, 0, 0	$4a/3$	$2\sqrt{2}/3 \approx 0.94$
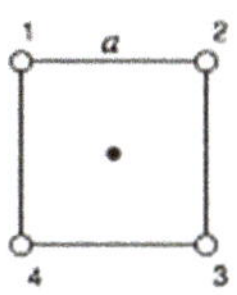	$-a, 0, a, 0$	$a, -a, a, -a$	a	$1/\sqrt{2} \approx 0.71$
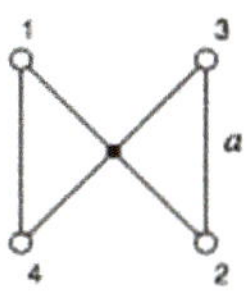	$-a, 0, a, 0$	$a, -a, a, -a$	$(1 + \sqrt{2})a/2$	≈ 0.92
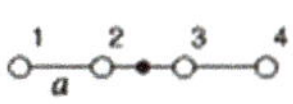	$-a, -a, -a, 3a$	0, 0, 0, 0	$3a/2$	$1/\sqrt{2} \approx 0.71$
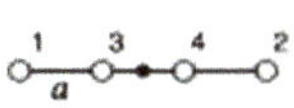	$-3a, 2a, -a, 2a$	0, 0, 0, 0	$2a$	1

The chosen viewing angles affect both the presence of ghost particles and the error associated with the 3D coordinate estimate. Graff and Gharib [5] performed a generalized analysis to access the error in the reconstructed scene (reconstruction precision) as well as the ghost particle generations (reconstruction quality) of various camera arrangements. Table 4.1 shows the error factor, E, defined as

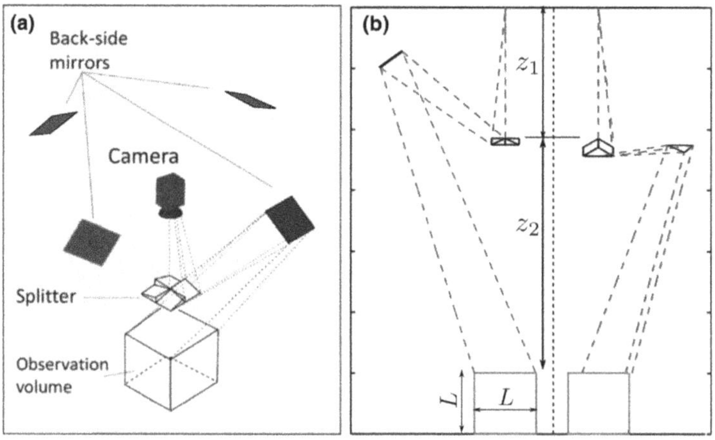

Figure 4.3. Experimental set-up of the single-camera 3D-PTV system. Reprinted from [6] with permission of Springer.

$$E = \sqrt{\sum_{i=1}^{N}\left[\left(\frac{\partial \bar{b}}{\partial x_i}\right)^2 + \left(\frac{\partial \bar{b}}{\partial y_i}\right)^2\right]}, \tag{4.3}$$

for N cameras with a mean image separation, $\bar{b}$. The mean image separation as well as its derivatives, $\frac{\partial \bar{b}}{\partial x_i}$, are dependent not only on their various camera arrangements but also on the order in which the identified particle images are processed between sequential cameras, assuming that each camera has the same parameters and uncorrelated error. The sequentiality of the particle images is identified by the numbering of the cameras shown in table 4.1, under the 'Set-up' column. The lowest error factor for a three-camera system is ~0.94, while that of a four-camera system is ~0.71, showing that a proper four-camera arrangement can perform better than a three-camera system. They also show that random ghost particles, which occur due to random chance that the point distribution generates false matches within the matching tolerance, are decreased with increasing number of cameras. However, clumped ghost particles, which occur when one or more point images are within a matching tolerance of another in a single-camera image, increase rapidly with increasing cameras, though they are more readily identifiable and easy to remove.

An alternative cost-effective solution for a 3D-PTV experimental set-up was developed by Kreizer and Liberzon [6]. The basic concept is to use a single camera with an image splitter and mirrors to obtain four different viewing angles of the observation volume. Figure 4.3 shows an example of this set-up. The authors also developed open source code based in Python to assist in the development of a similar system based on optical parameters[1].

[1] The software is available at http://www.openptv.net.

The accuracy of this experimental set-up, when calibrated properly, is the same as multi-camera 3D-PTV techniques. The drawback of this technique, however, is that each viewing angle uses one quarter of the camera's resolution.

4.2.1.2 Triangulation methods

Just as in 2D PTV, particle images are identified in the image plane of each camera individually, as discussed in section 4.1. From these 2D images, the 3D position of each object is reconstructed using triangulation, by first creating epipolar lines based on a pinhole camera model, then using their intersections to triangulate their 3D particle locations. This approach can be used for both stereoscopic (two-camera), or multi-camera (three or more cameras) PTV systems. For example, figure 4.4 [7] shows how triangulation works using epipolar lines for three cameras. E_{1-2} is the epipolar line created for pixel P' from image 1 into image 2. In this case, two cameras would not suffice for particle location, as there are three particle images in image 2 that lie on E_{1-2}. Thus the third camera is used, which would allow for generating epipolar lines, $E_{(2-3)i}$, associated with the three identified particles in image 2. It would also allow for generating an epipolar of P' from image 1 to image 3, E_{1-3}, the intersection of which would identify the correct triplets from which the particle's 3D spatial location can be determined. In reality, due to measurement uncertainties, a tolerance must be added about the epipolar lines to identify triplets within these uncertainties.

The mathematical formulation of photogrammetric PTV without any distortion is

$$x_i' = x_h - c\frac{r_{11}(X_i - X_0) + r_{21}(Y_i - Y_0) + r_{31}(Z_i - Z_0)}{r_{13}(X_i - X_0) + r_{23}(Y_i - Y_0) + r_{33}(Z_i - Z_0)}, \tag{4.4}$$

$$y_i' = y_h - c\frac{r_{12}(X_i - X_0) + r_{22}(Y_i - Y_0) + r_{32}(Z_i - Z_0)}{r_{13}(X_i - X_0) + r_{23}(Y_i - Y_0) + r_{33}(Z_i - Z_0)}. \tag{4.5}$$

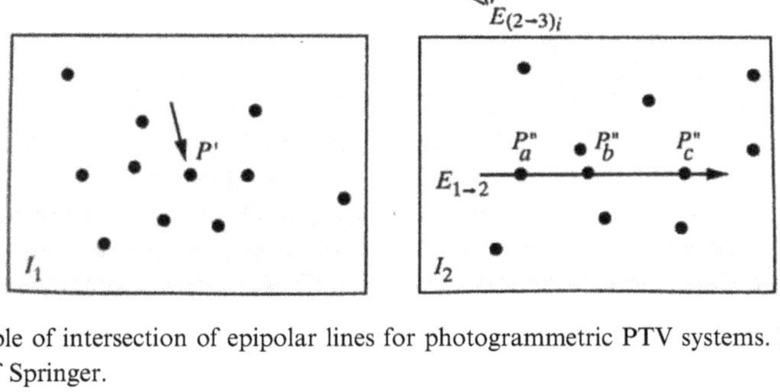

Figure 4.4. Principle of intersection of epipolar lines for photogrammetric PTV systems. Reprinted from [9] with permission of Springer.

In this case x_i' and y_i' are the image plane coordinates of a particle image and x_h and y_h are the coordinates of the center point in the image plane. The spatial coordinates are X_i, Y_i, and Z_i. X_0, Y_0, and Z_0 are the coordinates of the camera focal point and r is a 3×3 rotation matrix that captures the camera's external orientation.

To correct for lens distortion, Nishina *et al* [8] have suggested radial distortion corrections, however, Maas *et al* [9] have also included tangential distortion [10] corrections by modifying equations (4.4) and (4.5) as

$$\overline{x_i} = x_i' + dx_i \; \overline{y_i} = y_i' + dy_i \tag{4.6}$$

$$dx_i = x_i'\left(k_1 r_i'^2 + k_2 r_i'^4 + k_3 r_i'^6\right) + p_1\left(r_i'^2 + 2x_i'^2\right) + 2p_2 x_i' y_i' \tag{4.7}$$

$$dy_i = y_i'\left(k_1 r_i'^2 + k_2 r_i'^4 + k_3 r_i'^6\right) + p_1\left(r_i'^2 + 2y_i'^2\right) + 2p_1 x_i' y_i' \tag{4.8}$$

$$r_i'^2 = x_i'^2 + y_i'^2, \tag{4.9}$$

where k_1, k_2, k_3 are the radial distortions, and p_1, p_2 are the tangential distortions. Maas also proposed compensating for the effects of image digitization and storage using additional affine transformations as proposed by El-Hakim [11] and Albertz and Kreiling [12]. Calibration can be used to obtain these coefficients and is further discussed in section 4.3.3.

Maas *et al* [9] found that their implementation of photogrammetric PTV tracked on the order of 10^3 tracer particles before ghost particles began to appear. These particles were imaged within a $200 \times 160 \times 50$ mm^3 measurement volume, resulting in 0.0038 N_{ppp}. Particle positions were estimated to have errors of 60 μm in-plane and 180 μm out-of-plane. The RMS random errors for displacement were significantly smaller than for position, as the systematic errors in position estimates do not affect the accuracy of displacement estimates.

One method for increasing the particle seeding density, previously discussed in section 3.2.3, was developed by Lei *et al* [13], who developed a method to identify overlapped particles, which can then be combined with triangulation methods to obtain 3D-PTV results [14–16]. Another algorithm was developed by Fuchs *et al* [17] The first step of the algorithm reconstructs the volume using the tomographic multiplicative algebraic reconstruction technique (MART), described in section 4.2.2. The reconstructed volume is then binarized and the image particle locations are estimated from their center of mass. These images are then mapped back onto each camera sensor. They are then compared to the actual particle images. The true particle images first must be identified and located within each camera sensor using a 2D Gaussian fit. Matching is done between the reconstructed and true particle images. If a particle image is associated with multiple reconstructed images, it is disregarded. This reduces the effect of ghost particles. Triangulation is then performed for particle location if uniquely matching particle images exist on at least two sensors.

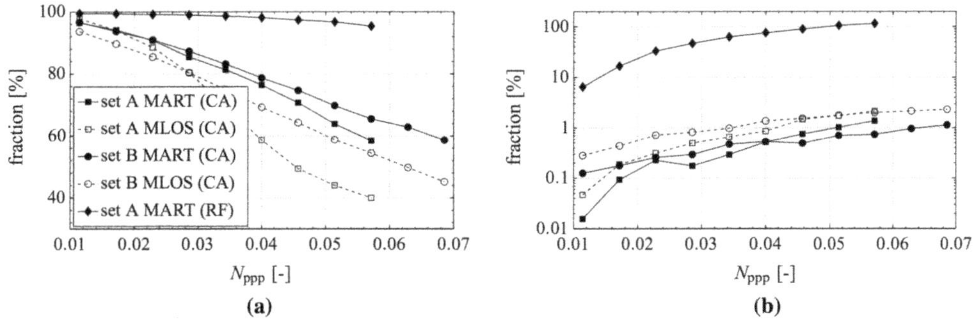

Figure 4.5. (a) The fraction of correctly estimated particle locations (within one voxel radius from the true particle location) and (b) the fraction of ghost particles compared to the number of true particles. Reprinted from [17] with permission from Springer.

The results of this reconstruction technique (CA) are compared to a standard tomographic reconstruction (RF) for two volume thicknesses (see figure 4.5). Set A has a volume size of $800 \times 800 \times 300$ voxel^3 and set B has a volume size of $800 \times 800 \times 50$ voxel^3. For example, at 0.05 N_{ppp}, the tomographic reconstruction (RF) shows near 100% correct particle identification, but also 100% generation of ghost particles. At 0.05 N_{ppp}, the proposed algorithm (CA) reduces the total number of correct particle locations to 70% (resulting in an effective 0.035 N_{ppp}), while also substantially reducing the fraction of ghost particles present in the reconstruction.

Most recently, Wieneke [18] introduced an iterative particle reconstruction (IPR) technique for performing triangulation that works well for particle seeding densities up to 0.05 N_{ppp}. The technique begins just the same as other triangulation methods: identification of particle images in 2D (although which method was used was not specified) and triangulation to determine 3D coordinates. The algorithm reconstructs particles in 3D and removes the images from each camera, allowing for further iterations of triangulation. After the initial 3D triangulation of the position and intensity of a particle p, $I^i_{\text{part}}(x_i, y_i, p)$, are calculated, a projection onto camera i, I^i_{proj}, is calculated as

$$I^i_{\text{proj}}(x_i, y_i) = \sum_p I^i_{\text{part}}(x_i, y_i, p). \tag{4.10}$$

The residual is then calculated as the difference of the projection and the original image. A further step calculates the residual for a particle image center shifted ±0.1 voxels in each direction. The particle location with the smallest residual is then selected. An intensity correction is then applied, and the residual image is then recalculated. Images are removed from an image if the intensity falls below a threshold or if it is located within one pixel of another image. Due to the removal of nearby particle images, IPR only requires three of four cameras to have corresponding intensity peaks to locate a particle in 3D.

Figure 4.6 shows the flow diagram for iterative particle reconstruction. There is a large loop that triangulates new particle images and their projections. There is also an inner loop that corrects particle locations and calculates the residual images. This technique is highly iterative and the computational time was on the same order of

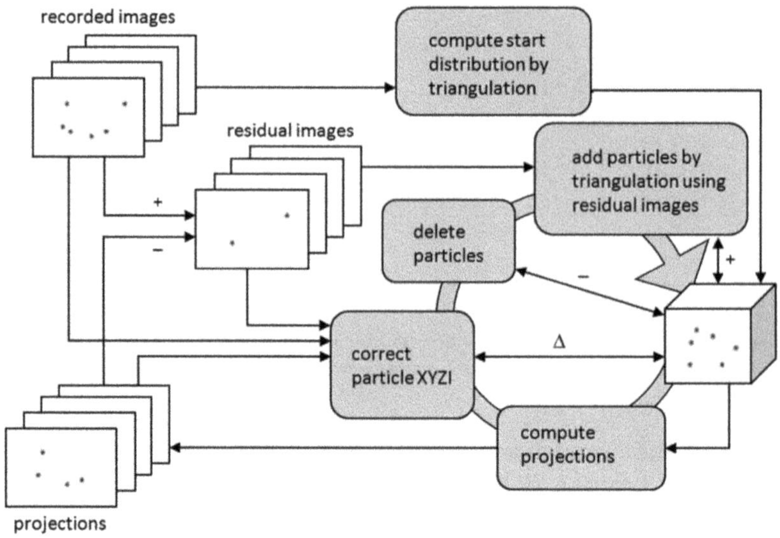

Figure 4.6. Flow diagram for iterative particle reconstruction [18].

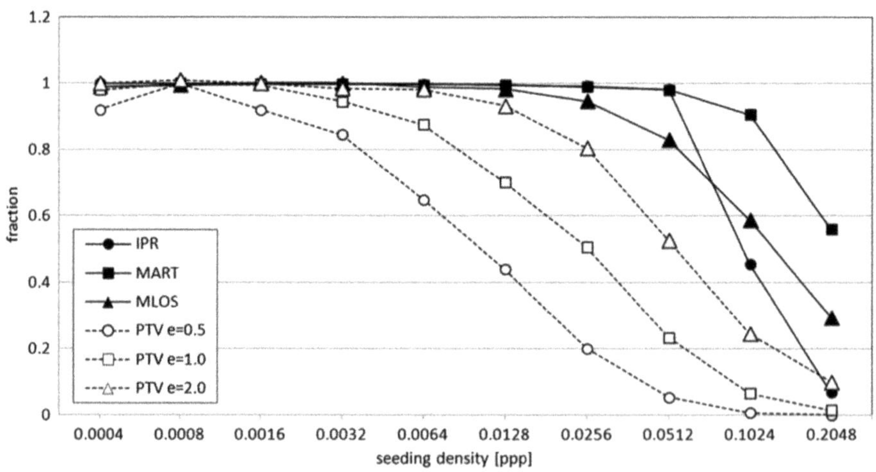

Figure 4.7. Fraction of detected true particles versus seeding density for IPR and other 3D reconstruction algorithms [18].

magnitude as MART reconstruction (see section 4.2.2). Despite the high computational cost, the algorithm is an improvement over previous 3D-PTV for high-density images. Figure 4.7 shows that IPR can identify as many particles as MART and more than 3D-PTV, which as stated is the same as a single-pass IPR, for seeding densities up to 0.05 N_{ppp}. Beyond this limit, the algorithm's performance drops off. Not only does the algorithm provide a high yield of true particle coordinates, but it also gives better accuracy than the other algorithms up to 0.05 N_{ppp}, as shown in figure 4.8.

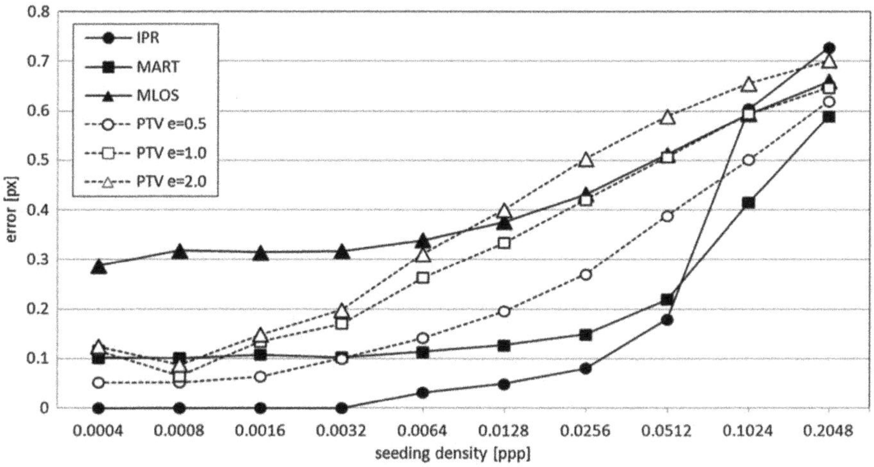

Figure 4.8. Mean 3D localization error versus particle seeding density for IPR and other 3D reconstruction algorithms [18].

4.2.1.3 Stereoscopic methods

Stereoscopic PTV (stereo-PTV) uses only two cameras to determine three velocity components. While stereo-PIV is widely used for planar measurements, a two-camera PTV set-up is not particularly common. This is due to the fact that PTV requires accurate and reliable determination of particle coordinates, which is difficult using only two cameras, in particular as N_{ppp} increases. Applications of stereo-PTV have therefore been few.

The first use of stereo-PTV was in a microscale experiment using two cameras and a stereomicroscope. Bown *et al* [19] then Yu *et al* [20] developed micro-stereo-PTV algorithms using different tracking techniques. Stereo-microscopes are designed to allow two cameras to image a single volume at microscales. There are two configurations that can be used: a Greenough type and a common main objective type (CMO). These configurations are shown in figure 4.9, where the CMO type gives a translational configuration of stereo imaging and the Greenough type gives an angular displacement configuration. The CMO microscope offers the advantage of sharpness throughout the entire field-of view and smaller bias and RMS errors than the Greenough configuration (see figure 4.10); however, the image generated using a CMO configuration is axially asymmetrical because the beams do not pass through the center of the objective lens. Therefore, the optical aberrations are easier to correct for in the Greenough type microscope.

In order to determine 3D coordinates using two cameras, a calibration procedure is performed to determine a direct linear transform (DLT) for conversion between image and object spaces. This is a general mapping function,

$$X = F(x), \tag{4.11}$$

where F is a third-order polynomial that relates image coordinates X to the object coordinates x. In order to determine object coordinates, stereoscopic correspondences

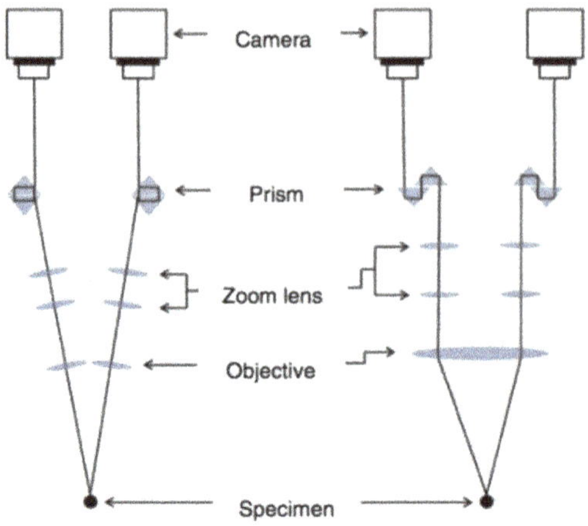

Figure 4.9. Schematics of stereoscope configurations with CMO (translation) and Greenough (angular displacement) type microscope systems [20].

must be determined. Bown *et al* [19] suggested that a cross correlation should be performed on the left and right images to estimate the image shift between the two image spaces. For each particle in the first image space, a correspondence is considered valid if exactly one particle in the second image space lies within a specified tolerance (set to 10 pixels) of the estimated image shift. Stereoscopic pairs are then transformed to object coordinates using the F matrix.

Bown *et al* [19] performed an experiment using a backward-facing step flow with a field-of view of 900 × 720 μm and a displacement on the order of 10 pixels, the particle tracks were averaged over 100 images. 2D averaging was performed on a $45 \times 45\ \mu m^2$ grid to be compared to stereo-PIV and 3D averaging was done across $10 \times 10 \times 10\ \mu m^3$ with a depth of 45 μm. Comparing these average values to DNS solutions, the 2D PTV results had uncertainties of 2% of mean velocity for in-plane flow and 3% of mean velocity for out-of-plane flow. This is an improvement compared to 3% in-plane and 7% out-of-plane uncertainties for PIV. 3D averaging results gave experimental uncertainties of 2% of the mean velocity for in-plane components and 5% of the mean velocity for out-of-plane components. The 3D averaging gives a full three-dimensional, three-component mapping of the flow field. This averaging method is only applicable for steady flows.

Yu *et al* [20] took a different approach for low seeding densities that finds particle tracks in each image individually using a relaxation method (see section 5.3). The particle positions are then transformed into the object plane in physical coordinates. Particle tracks having the shortest distance in the projections between left and right images are then selected as stereoscopic matches. Using the two corresponding 2D displacements in the left and right images, the stereoscopic approach for determining 3D displacements is then used, where a transformation matrix, F, is approximated for each individual particle pair as ∇F:

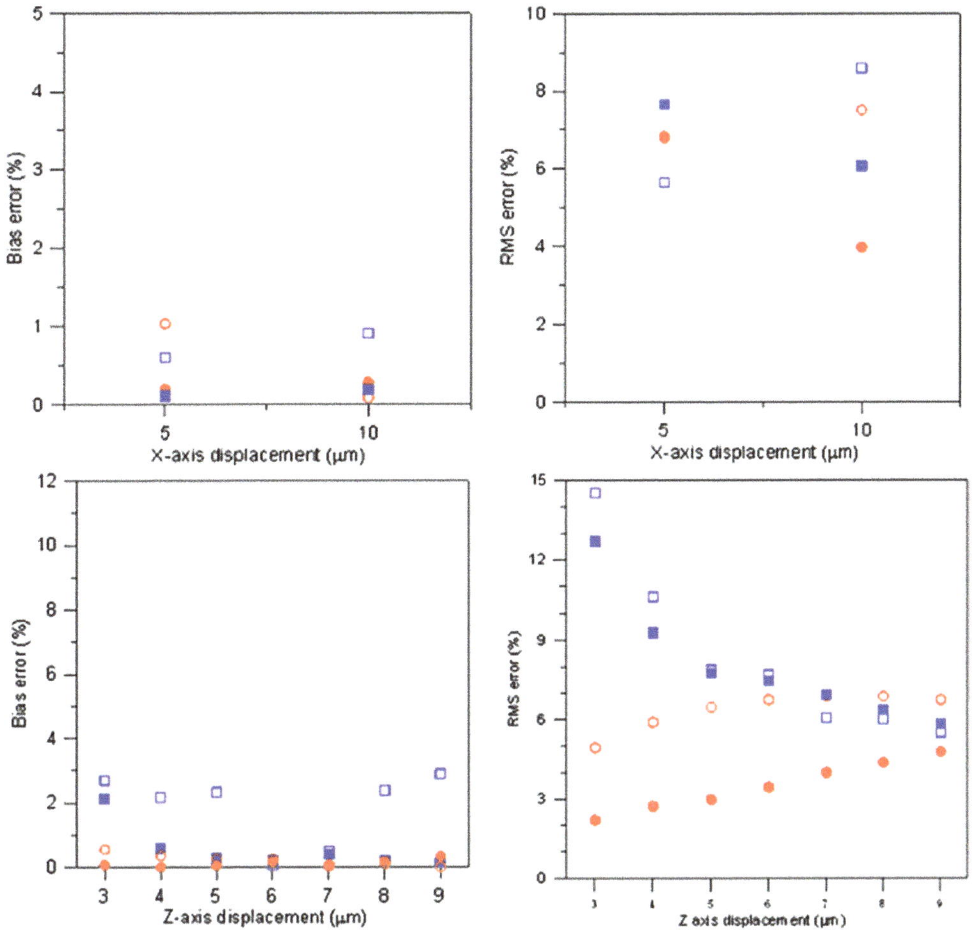

Figure 4.10. Bias and RMS errors in the x and z directions using stereo-PTV (circles) and stereo-PIV (squares) with Greenough type (outlined) and CMO type (filled). Reprinted from [20] with permission of Springer.

$$\begin{bmatrix} \Delta X_l \\ \Delta Y_l \\ \Delta X_r \\ \Delta Y_r \end{bmatrix} = \begin{bmatrix} F_{xl,\,X} & F_{xl,\,Y} & F_{xl,\,Z} \\ F_{yl,\,X} & F_{yl,\,Y} & F_{yl,\,Z} \\ F_{xr,\,X} & F_{xr,\,Y} & F_{xr,\,Z} \\ F_{yr,\,X} & F_{yr,\,Y} & F_{yr,\,Z} \end{bmatrix} \begin{bmatrix} \Delta x \\ \Delta y \\ \Delta z \end{bmatrix}. \tag{4.12}$$

This gives three components of velocity; however, it does not assign a depth, thus this technique is a two-dimensional three-component (2D3C) PTV technique.

In order to test the accuracy of this stereoscopic tracking algorithm, an experiment was performed by translating a known calibration target in the x and z directions. This was done using both cross-correlation stereo-PIV and the stereo-PTV algorithm used by Yu *et al* [20], with both a Greenough type and a CMO type stereomicroscope. The results are presented in figure 4.10, where the bias errors

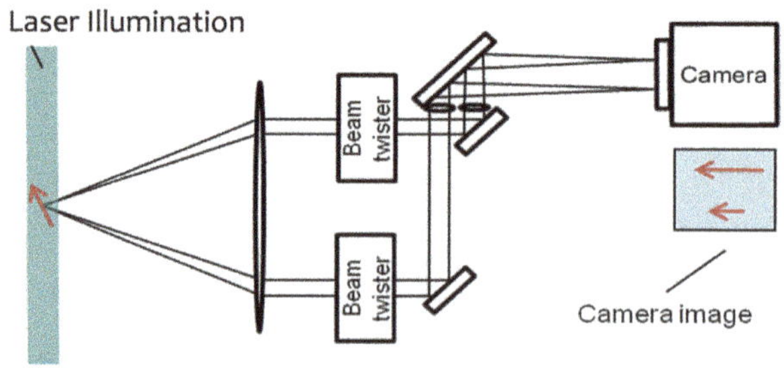

Figure 4.11. A single camera using two off-center optical paths through a single lens. Reprinted from [21].

associated with PIV techniques were reduced for PTV both in-plane (x direction) and out-of-plane (z direction). Both the x and z direction bias errors were less than 1% of the true displacement for PTV, although the RMS errors ranged from 4% to 8% in the x direction and 2% to 7% in the z direction. In the majority of tested cases, the stereo-PTV algorithm using the CMO type stereomicroscope had the lowest errors.

Peterson *et al* [21] made a modification to the microscopic stereo-PTV algorithms presented in order to use a single camera. Figure 4.11 shows how a camera's view was split in two in order to record two images in the same frame. Both views were reoriented by 90 degrees by beam twister mirror set-ups, which made more efficient use of the image sensor space.

Peterson *et al* [21] used the concept of epipolar lines, as described in section 4.2.1.2, with an additional constraint on image-similarity. Stereoscopic pairs are determined by minimizing the difference between epipolar lines in 3D space. If a pair consists of only particles that are not claimed by another pair, then they are considered a match. Otherwise, the pair with the highest degree of similarity is chosen as valid. The seven proposed similarity criteria are: peak intensity, summed intensity, total number of pixels, image width, image height, maximum value of a cross-correlation between raw particle images, and maximum value of a cross-correlation between binarized particle images. No single similarity criteria was capable of separating valid pairs, so thresholds were determined for all seven criteria such that each allowed for 90% of valid sets to be recognized. The proposed algorithm was applied to a flow exiting a hose and compared to PIV results. The two algorithms agreed within 9% of the velocity measurements.

Bao and Li [22] proposed a method for stereoscopic particle image pairing that utilizes a combination of photogrammetric and binocular vision principles. In binocular vision, a point source of light that is imaged outside of the focus plane will appear as a circle. The depth of the point source from the camera lens can be estimated from the radius of this blur circle r as

$$D = \frac{Fv_0}{v_0 - F - rkf}, \tag{4.13}$$

where F is the focal length of a thin convex lens, v_0 is the distance from the lens to the image plane, and f is the f-number of the camera lens. The parameter k needs to be determined from a calibration procedure. Figure 4.12 shows the geometry of this binocular vision set-up and shows how an out-of-focus particle will appear as a circle in the image plane.

The procedure for stereoscopic pairing proposed by Bao and Li [22] began by selecting an unpaired reference particle, p_0, in the first image. The epipolar line segment with a depth range obtained from the blur circle radius and its uncertainty is then calculated. The candidate particle, $p_{\min}'$, in the second image that is closest to this line segment is then selected. If this particle is within the depth estimate of p_0, then it is considered the optimal pair particle. If it is not, then $p_{\min}'$ is labeled as an outlier and the next closest particle is considered as a candidate. After a valid particle is selected, this procedure is repeated for each reference particle in the first image. This technique was shown to increase the correct pairing rate for a simulation with 2200 particle pairs (0.0011 N_{ppp}) from 56.1% to 74.3% when compared to an epipolar line nearest neighbor analysis. This technique showed that stereoscopic correspondences can be determined in a 40 mm thick volume using only two cameras.

Panday *et al* [23] introduced a technique for 3D localization with two cameras that uses an epipolar line search and ant colony optimization. Ant colony optimization (ACO) is based on the methodology of an ant colony when searching the surrounding area for food. If a particular path is considered good by the ant, it will leave behind some amount of pheromone for other ants. In the context of stereo-PTV, a path chosen by an ant is from one particle image from camera 1 to another particle image in camera 2. The path between particle images is considered good if the distance between epipolar lines is short. Ants make decisions about which path to follow based on pheromone amounts and local information, characterized by equation (4.15), where p^k is the probability an ant takes the path from particle i to particle j, τ is the pheromone measure between the two points, η is the reciprocal of the distance between particles i and j, and α and β are the parameters that determine the importance of local and global information,

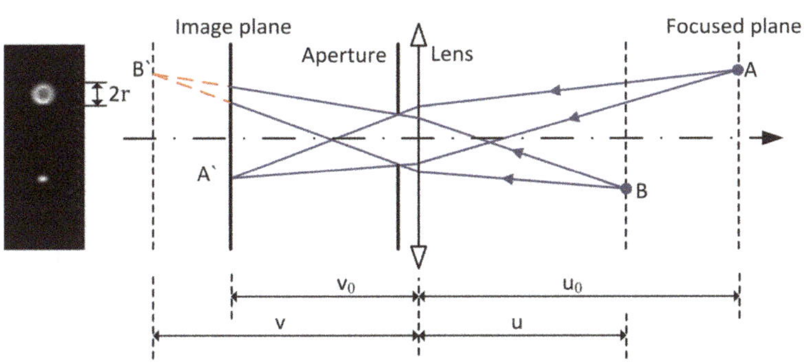

Figure 4.12. Geometry of defocused image formation [22].

$$P^k(i, j) = \frac{[\tau(i, j)]^\alpha [\eta(i, j)]^\beta}{\sum_{l \in N^k} [\tau(i, l)]^\alpha [\eta(i, l)]^\beta}. \tag{4.14}$$

Ants visit every particle, i, in the left frame and decide which particle in the right frame to visit. After every ant visits every particle in its itinerary, pheromone amounts are updated. Equation (4.15) gives the update procedure for the pheromone left by an ant between points i and j, and ρ is the pheromone decay rate, which ensures there is no unlimited accumulation of pheromone:

$$\tau(i, j) \leftarrow (1 - \rho)\tau(i, j) + \sum_{k=1}^{m} \Delta\tau^k(i, j), \tag{4.15}$$

where

$$\Delta\tau^k(i, j) = \begin{cases} \frac{1}{L^k} & \text{if } (i, j) \in \ T^k \\ 0 & \text{otherwise} \end{cases}, \tag{4.16}$$

where T_k is the set of portions of route included in the itinerary of ant k and L^k is the total distance of the itinerary route. Thus, the change in pheromone amount is determined by how well an ant has chosen its route.

Two modifications to this ant system (AS) method were proposed. The first modification was named the ant colony system (ACS) method. It differed from AS in three aspects: the selection probability of an ant agent to go from one particle in the first image to the other particle in the second image, the global updating rule was only applied to routes belonging to the best ant tour, and local pheromone updating is applied as ants construct a solution. The new selection probability for an ant agent is

$$P*^k(i, j) = \frac{[\tau(i, j)][\eta(i, j)]^\beta}{\sum_{l \in N^k} [\tau(i, l)][\eta(i, l)]^\beta}, \tag{4.17}$$

$$\begin{cases} p^k(i, j) \ = \ \begin{cases} 1 \text{ if } j \ = \ \operatorname{argmax} p*^k(i, l) \\ \quad 0 \text{ otherwise} \end{cases} & \text{if } q \leqslant q_0 \\ p^k(i, j) = p*^k(i, j) & \text{if } q \ > \ q_0 \end{cases}, \tag{4.18}$$

where q is a random variable uniformly distributed over [0,1] and q_0 is a parameter between zero and one that determines the relative importance of exploitation versus exploration. If exploration is chosen ($q \leqslant q_0$), then the ant makes a decision based on available information. If exploitation is chosen ($q > q_0$), then biased exploration is performed as in the AS method.

The pheromone update in ACS is done in two steps. The first step takes place after each ant travels from a single path from particle i to j. This updated is

$$\tau(i, j) \leftarrow (1 - \psi)\tau(i, j) + \psi\tau_0, \tag{4.19}$$

where ψ is the local evaporation rate of the pheromone. This update causes paths that are selected to have a reduced pheromone so that other ants are persuaded to choose a different path in an iteration. This keeps ants from converging to a common path. The second step of the pheromone update takes place after all ant agents complete their itinerary

$$\tau(i, j) \leftarrow (1 - \rho)\tau(i, j) + \rho\Delta\tau(i, j), \tag{4.20}$$

where

$$\Delta\tau(i, j) = \begin{cases} \dfrac{1}{L^+} & \text{if } (i, j) \in \ T^+ \\ 0 & \text{otherwise} \end{cases}, \tag{4.21}$$

where T^+ is the best particle link route ever traveled by all the agents and L^+ is the total distance of that best route. Thus, only the best route is used to update pheromones.

The second modification of the AS algorithm is based on Q-learning properties. This algorithm is called Ant-Q and differs from ACS only in the first step of the local pheromone updating rule, which is given by

$$\tau(i, j) \leftarrow ((1 - \psi)\tau(i, j) + \psi\gamma * \max_{l \in N^k} \tau(j, l), \tag{4.22}$$

where γ is a new parameter between zero and one. The idea behind this modification is to update local pheromone trails with a value, which was a prediction of the value of the next state.

Panday *et al* [23] tested these algorithms using stereoscopic PIV standard images with a resolution of 256 × 256 pixels. The parameters used in this study were: α and β were 1.0 and 5.0, respectively, ρ was 0.5, τ_0 was 1.0, ψ was 0.5, and γ was 1. Five iterations were done using 20 ant agents. The correct matching rates for each ant colony algorithm compared to a nearest neighbor epipolar line search is shown for standard images with a range of seeding densities from 0.0024–0.024 N_{ppp} in table 4.2. Ant colony pairing improved the performance of stereo matching over the entire range of particle densities tested. The limit of the algorithm was a correct pairing rate of 95% for Ant-Q with a seeding density of 0.0024 N_{ppp}. The authors suggested that increased accuracy would require more cameras.

4.2.1.4 Scanning 3D-PTV

An alternative technique for capturing 3D velocity fields is the process of scanning multiple planes separately, introduced by Hoyer *et al* [24]. The technique typically uses one or two cameras as opposed to the four to six used in other 3D methods. Figure 4.13 gives an example experimental set-up. Scanning PTV records multiple frames to reconstruct a single volumetric measurement; therefore, in order to compare to other 3D-PTV techniques, scanning PTV (SPTV) must be able to scan an entire volume in the same time that a single frame is recorded otherwise. For this reason, the frame rate of SPTV must be increased by a factor equal to the number of slices of flow measured over the frame rate of 3D-PTV, which is limiting

Table 4.2. Ant colony optimization results for stereo image pairing. [23].

		Correct pair rate (%)			
Series #/Frame #	Existing pairs	Conventional	AS	ACS	Ant-Q
#351/000	1546	68.69	72.06	71.6	72.12
#351/001	1526	70.77	71.75	72.02	72.08
#352/000	283	88.69	91.52	93.64	92.93
#352/001	277	88.09	92.77	92.42	92.78
#371/000	157	89.17	91.08	93.63	93.63
#371/001	160	89.38	95	93.75	95
#377/000	352	75	77.56	76.42	75.85
#377/001	352	76.42	80.11	82.1	81.82

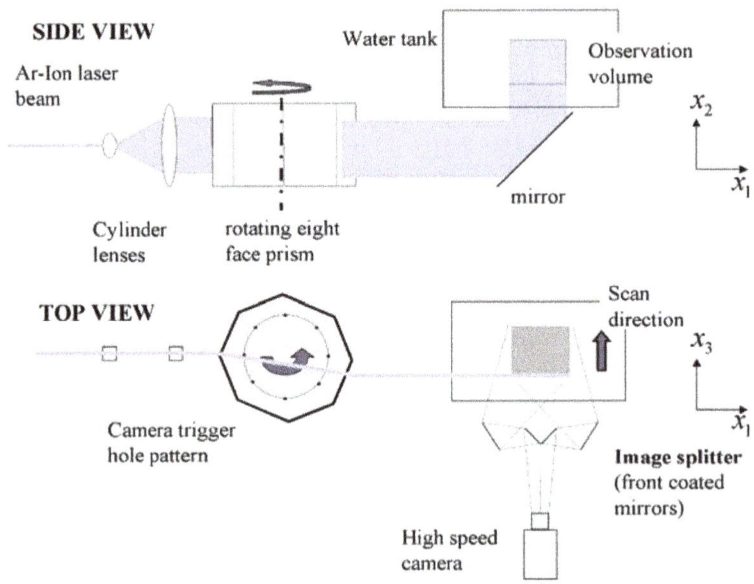

Figure 4.13. Experimental set-up with a laser beam being expanded to a sheet and scanned through a volume using a rotating eight face prism. Reprinted from [24], Copyright (2011), with permission from Elsevier.

in the context of time-resolved measurement. The benefit of scanning is that a single-camera frame only contains particle images from a thin cross-section of the observation volume, thereby removing ghosts from the triangulation process.

Hoyer *et al* [24] used single camera with an image splitter to obtain four views of the observation volume. The four views were used for triangulation, as done by Maas *et al* [9], to determine precise 3D particle coordinates within each plane. When scanning a volume in SPTV, two phenomena need to be accounted for: frames are captured at skewed points in time and particles can move between slices between volume scans. When searching for a particle's track, Hoyer *et al* [24] introduced the constraint that a particle can be expected to be found in either the same plane or an adjacent plane between two volume scans. This requires an estimate for the maximum out-of-plane flow velocity as well as an appropriate frame rate. When these were accounted for, Hoyer *et al* [24] found that they were able to double both the observation volume and the spatial resolution compared to its contemporary photogrammetric 3D-PTV. The SPTV identified on average 3900 particles per volume scan, which is equivalent to 0.0037 N_{ppp} for the single 1024 × 1024 pixel camera that was used. Due to their hardware limitations, the camera memory could only record data for 4 s at a time, thereby limiting the ability of their SPTV system for statistical studies. With hardware improvements, however, this can be overcome.

4.2.1.5 Defocusing methods

Willert and Gharib [25] suggested a photogrammetric method based on the same principles but only using one camera. Called the defocusing DPIV (DDPIV) technique, it uses a three-hole aperture, where the holes are placed at the vertices of an equilateral triangle centered about the lens' optical axis, to produce three images for each particle. Based on the separation of the images, the distance from a reference plane can be determined. This concept is shown in figure 4.14, where imaging with a single on-axis pinhole is compared with imaging with two off-axis pinholes.

In both cases, it can be seen that when a particle is on the reference plane, particle *A* images sharply onto the CCD sensor. In the former case, however, particle *B* focuses behind the CCD, resulting in a slightly blurred image on the CCD sensor. In

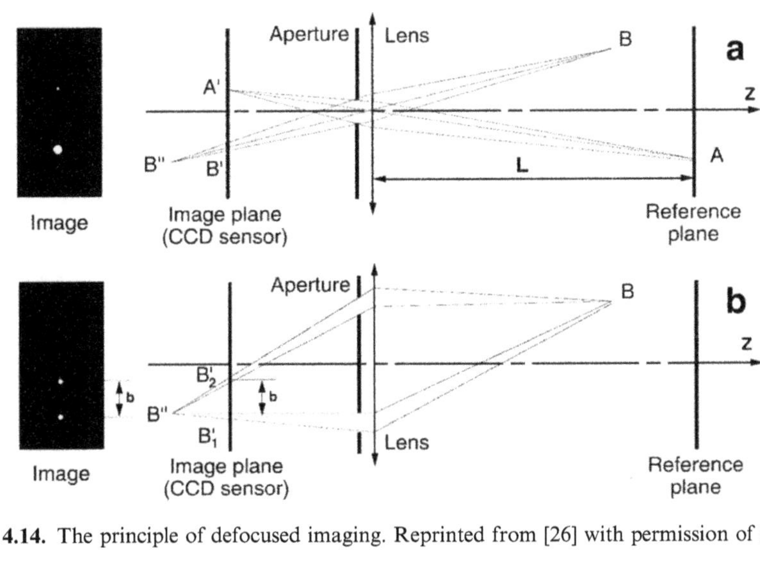

Figure 4.14. The principle of defocused imaging. Reprinted from [26] with permission of Springer.

the latter, case, particle B still focuses behind the CCD sensor. However, because it images through the two off-axis pinholes, it results in two slightly blurred images onto the CCD sensor. In this manner, the separation between the particle's images, b, correlates with the particle's depth location, while the pair's lateral location correlates to the particle's planar position. Using this two-pinhole arrangement, Willert and Gharib [25] derived the relations expressing a particle's spatial location as

$$X = \frac{-x_0 Z}{ML}, \tag{4.23}$$

$$Y = \frac{-y_0 Z}{ML}, \tag{4.24}$$

$$Z = \frac{1}{\frac{1}{L} + Kb}, \tag{4.25}$$

$$K = \frac{1}{MLd}, \qquad b = \frac{Md}{Z}(L - Z), \qquad M = \frac{f}{L - f}, \tag{4.26}$$

where (x_o, y_o) are the geometric centers of the particle images' locations, d is the pinhole separation, L is the distance from the aperture to the reference plane, and M is the geometric magnification. The sensitivity of the system to detect depth changes is given as

$$\frac{\partial b}{\partial Z} = -\frac{1}{KZ^2}. \tag{4.27}$$

Once particle images were identified in the CCD images; these equations were used to obtain the particles' 3D spatial positions.

Yoon and Kim [27] noted that in the single lens microscopic applications, the distance between the pinhole mask and the lens plane is not negligibly small, and that the misalignment error of the pinhole mask with the lens is also not trivially small, thereby introducing noticeable errors. Towards this end, they proposed an x–y compensation calibration procedure using polynomial calibration functions to correct for the x and y positions of the centers of the particles on the image plane. Using 17 particles on their target for this calibration procedure, they obtained uncertainties of 0.090/0.098/0.26 μm for the x/y/z positions, respectively, across a 50 μm volume depth. With this approach, they were able to obtain velocity measurements over a micro backward-facing step with dimensions 768 μm long by 388 μm wide by 50 μm deep, although no particle image density was reported. Pereira *et al* [28] suggested using a calibration procedure to obtain camera parameters, which entailed placing 2 μm fluorescent particles onto a transparent plate, and sequentially moving it through known depth locations and acquiring images. These were then used with equations (4.23)–(4.26) to identify 3D particle positions.

Because each particle is triply exposed onto the CCD, this technique is limited to low seeding. Tien *et al* [14, 15] suggested placing three different color filters over each pinhole and using a 3-CCD color camera. Using the color filters and a white backlight, particle images appear as cyan (green and blue), yellow (red and green), and magenta (blue and red) on a white background (see figure 4.15 left). Furthermore, Tien *et al* [15] recognized that color aberration results in different peak locations for different wavelengths (see figure 4.16) causing particle location errors, and resolved this with a new configuration by aligning red, green, and blue LED sources with each of the pinholes for illumination (see figure 4.17). In order to eliminate any crosstalk between the camera's color channels, a principal component transformation was implemented on the acquired images, creating new images based on the new independent primaries. The volume imaged was cubic and the mean and variance of error in an out-of-plane location are plotted in figure 4.18, where it can be seen that the mean errors are within 1.5%, the x and y standard deviation errors are within 0.04%, and the z standard deviation errors are within 0.25%. To further improve this approach, Tien [29] suggests using two dichroic beam splitters to

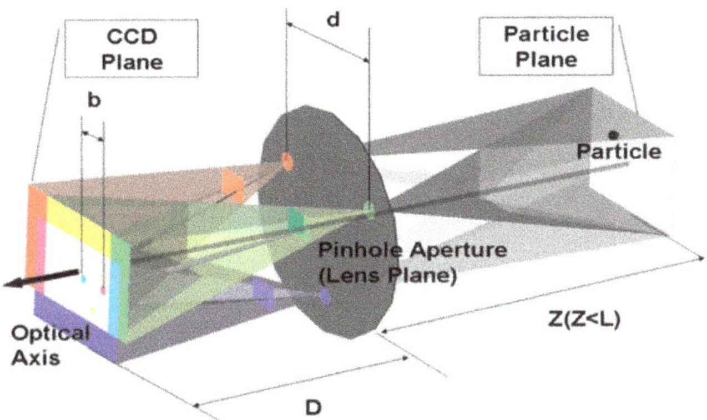

Figure 4.15. Defocused images obtained from a black particle illuminated with a white backlight and three color filters. Reprinted from [14, 15] .

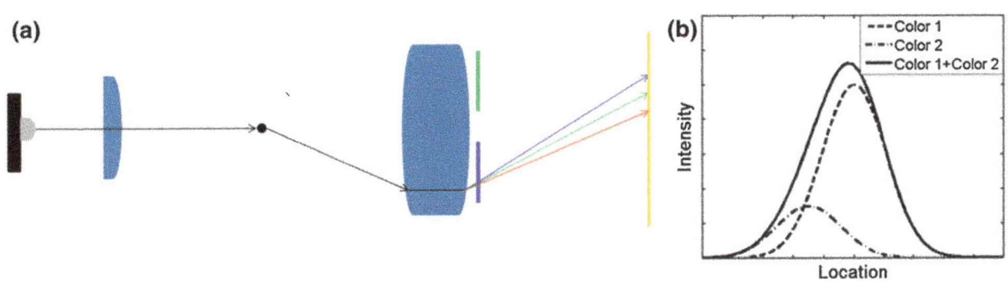

Figure 4.16. Particle image peak shifts due to color aberration: (a) color aberration due to white light illumination and (b) the skewed Gaussian distribution. Reprinted from [15] with permission of Springer.

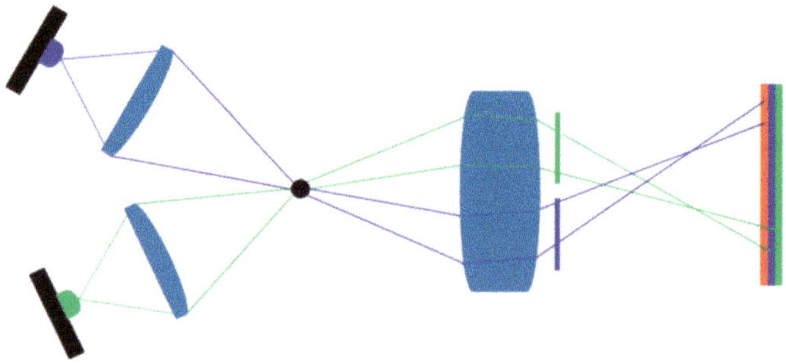

Figure 4.17. Improved illumination configuration. Reprinted from [15] with permission of Springer.

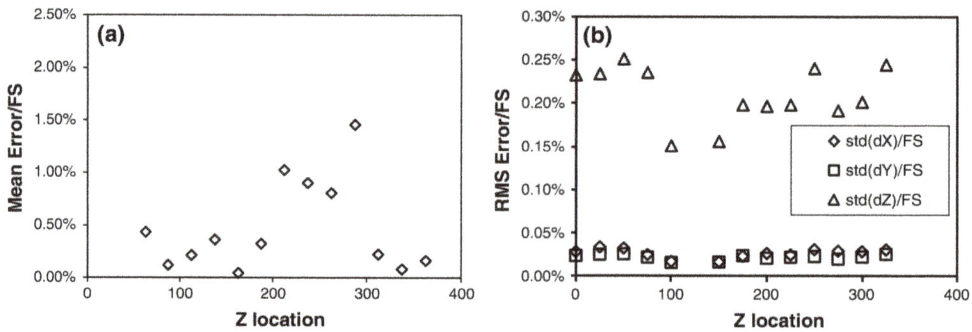

Figure 4.18. Out-of-plane location error as a percentage of the calibration range. Reprinted from [15] with permission of Springer.

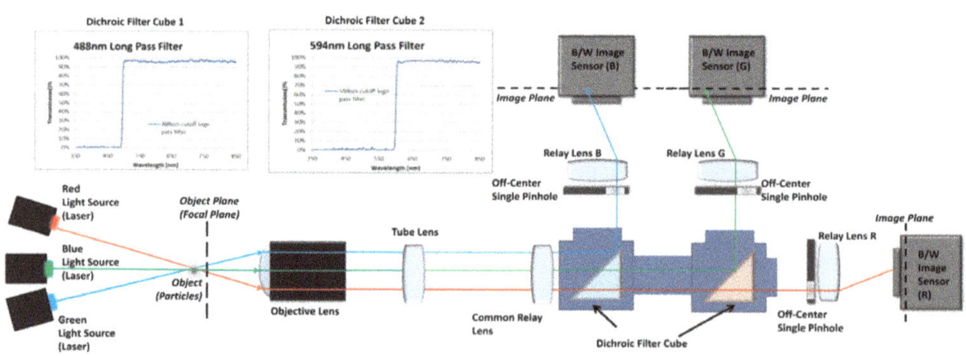

Figure 4.19. Concept of the proposed multi-spectral 3DμPTV system [29].

separate the incoming light into three spectral ranges, and to image each of these onto monochrome cameras (see figure 4.19).

An alternative implementation of DDPIV uses three separate cameras. In this set-up, each CCD is arranged around a common system axis, such that the axis of each camera and lens is radially tilted towards the common system axis so that they

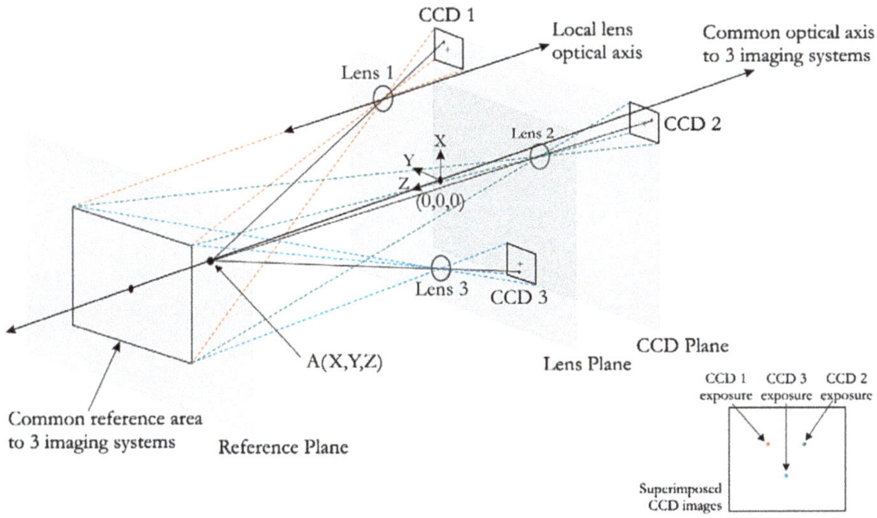

Figure 4.20. Implementation of 3D defocusing PIV [30].

all intersect at the same point on the system axis [30, 31]. This multi-camera arrangement, shown in figure 4.20 [33], is equivalent to the single-camera set-up shown in figure 4.15, aside from the fact that each particle will only create one image on each CCD, thereby eliminating in an alternative approach the problem of image overcrowding associated with the initial DDPIV system.

Grothe and Dabiri [31] defined an improved volume of interest by taking into account the dimensions of the CCDs. The new definition changes the imaging volume from a tetrahedron into a truncated rectangular pyramid. This was done by shifting the optical axis of the system to the half-height and half-width of the triangle, rather than its centroid. This allows for full exposure of the CCDs and a more practical rectangular, symmetric imaging volume. The comparison of camera set-ups and imaging volume are shown in figure 4.21 and figure 4.22, respectively. The performance of the modified set-up was shown to be similar to that of Kajitani and Dabiri. The absolute error ratio for the rectangular domain was shown to be 0.68% lower than that of the tetrahedral domain.

Pereira and Gharib [32] applied the technique for particle densities on the order of 0.024 N_{ppp}. For a three-pinhole system, Kajitani and Dabiri [30] defined γ as the distance from the pinholes' geometric center (the system's optical axis) to each of the pinholes, and ζ as the distance from the particle image's triplet geometric center to each of the particle's images, and derived a full three-dimensional characterization of the three-pinhole system proposed by Willert and Gharib [25]. Grothe and Dabiri [31] further improved on the full characterization of this system. Uncertainty analyses were performed for the three systems and are shown in table 4.3.

The error for $\delta(dY)$ shown in table 4.3 shows that this error is minimum on the system optical axis, however, it grows with lateral locations. Grothe and Dabiri [31] plotted their results as in figure 4.23, showing the error ratio of in-plane and

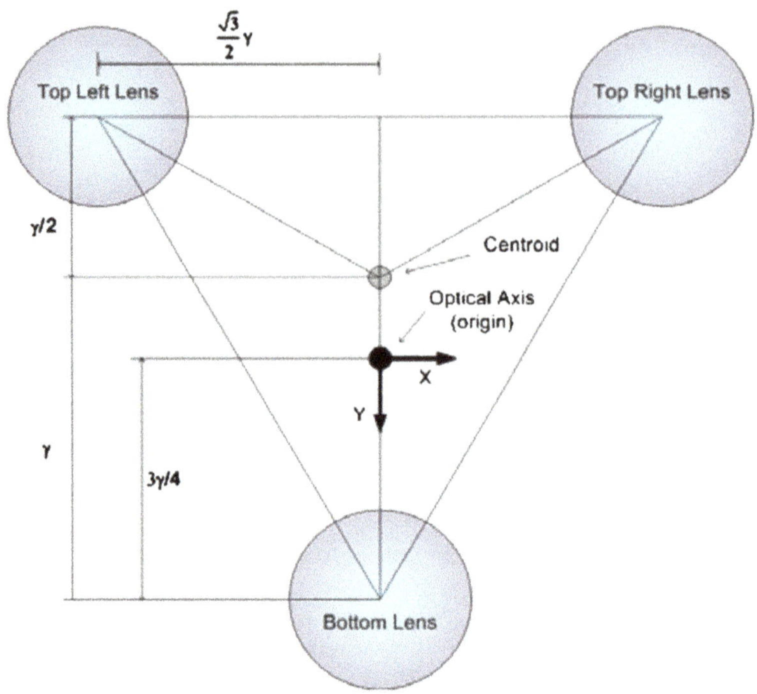

Figure 4.21. Location of the optical axis for a rectangular imaging domain [31].

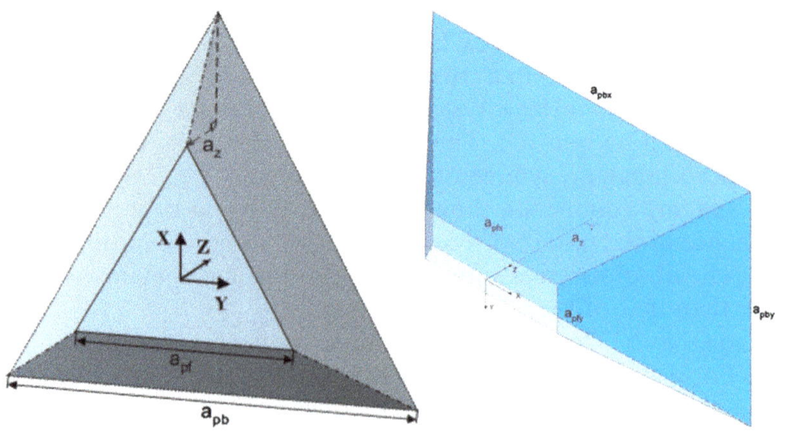

Figure 4.22. Tetrahedral (left) and rectangular (right) imaging domains for defocusing PIV [31].

out-of-plane measurements. Here, it can be seen that errors are dependent on the non-dimensional depth and off-axis position. The peak error ratio occurs at the system optical axis, and drops with lateral location per given depth location. This is due to increasing errors in $\delta(dY)$ for given Z locations and hence fixed errors in (dZ).

Table 4.3. Comparison between two-dimensional, Kajitani and Dabiri, and Grothe and Dabiri system uncertainties.

	2D derivation	Kajitani and Dabiri	Grothe and Dabiri
$\delta(dX)$	$K\lvert Z\rvert d\Delta x\sqrt{1+\frac{2X^2}{d^2}}$	$K\lvert Z\rvert\gamma\Delta x\sqrt{1+\frac{X^2}{\gamma^2}}$	$K\lvert Z\rvert\gamma\Delta x\sqrt{1+\frac{X^2}{\gamma^2}}$
$\delta(dY)$	$K\lvert Z\rvert d\frac{\Delta x}{\sqrt{2}}\sqrt{1+\frac{2Y^2}{d^2}}$	$\lvert Z\rvert\gamma\Delta x\sqrt{1+\frac{Y^2}{\gamma^2}}$	$K\lvert Z\rvert\gamma\Delta x\sqrt{17+8\frac{X}{\gamma}+16\frac{X^2}{\gamma^2}}$
$\delta(dZ)$	$\sqrt{2}\,KZ^2\Delta x$	$KZ^2\Delta x$	$KZ^2\Delta x$
$\delta(db)$	$\sqrt{2}\,\Delta x$	Δx	
$\frac{\delta(dZ)}{\delta(dY)}$	$\frac{2\lvert Z\rvert}{d\sqrt{1+\frac{2Y^2}{d^2}}}$	$\frac{\lvert Z\rvert/\gamma}{\sqrt{1+\frac{2Y^2}{\gamma^2}}}$	$\frac{4\lvert Z\rvert/\gamma}{\sqrt{17+8\frac{X}{\gamma}+16\frac{X^2}{\gamma^2}}}$

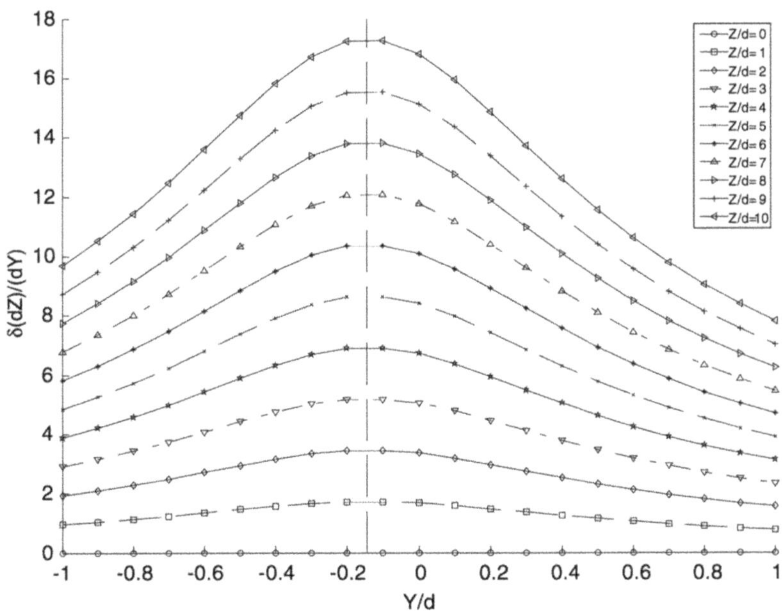

Figure 4.23. Error ratio of the out-of-plane to in-plane measurements. Z and X are both normalized with respect to d, the length of a side of the equilateral triangle aperture arrangement [33].

4.2.1.6 Anamorphic methods

The astigmatism PTV (APTV) technique [33–42] implements an anamorphic approach using a single camera and a cylindrical lens to stretch the particle images from circular to elliptical, such that depth information can be extracted from this deformation. While this method has been used in macroscopic scale experiments [34, 35, 40], it was first used in microscale experiments by Kao and Verkman [33], and has since predominantly been used in microscale experiments. Using the lens equations

$$\frac{1}{u_x} + \frac{1}{v_x} = \frac{1}{f} \,\&\, \frac{1}{u_y} + \frac{1}{v_y} = \frac{1}{f'}, \tag{4.28}$$

Chen *et al* [36] showed that the measurable range, Δu, of the system is

$$\Delta u \approx \left(\frac{f - d}{f \times f_x} + \frac{p_2}{(v_x + p_2) \times v_x}\right) u_m^2 - p_1, \tag{4.29}$$

where u_m is the distance from the center of the measurable range to the first principal plane of the combined lens, where u_x and u_y are the distances between the in-focus planes and the first principal plane of the combined lens in the x and y directions, respectively, v_x and v_y are the image distances in the x and y directions, respectively; f and f_c are the focal lengths of the field and cylindrical lenses, respectively; d is the distance between the two lenses; f' is the effective focal length of these two; p_1 and p_2 are the distances from the first and second principal planes of the combination of these two lenses to the field lens; and

$$p_1 = \frac{f \times d}{f + f_c - d}, \; p_2 = \frac{(f - d)d}{f + f_c - d}. \tag{4.30}$$

If the cylindrical lens is close to the field lens, where $d = 0$ and $p_1 = p_2 = 0$, then the measurement field reduces to

$$\Delta u \approx \frac{u_m^2}{f_c}, \tag{4.31}$$

showing that the measurable range is inversely proportional to the focal length of the cylindrical lens (figure 4.24). Figure 4.25 shows an example of the anamorphic

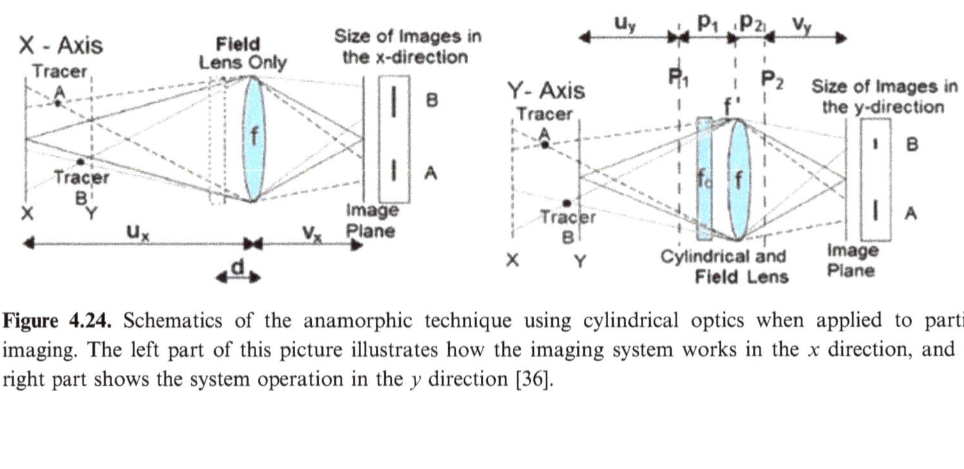

Figure 4.24. Schematics of the anamorphic technique using cylindrical optics when applied to particle imaging. The left part of this picture illustrates how the imaging system works in the x direction, and the right part shows the system operation in the y direction [36].

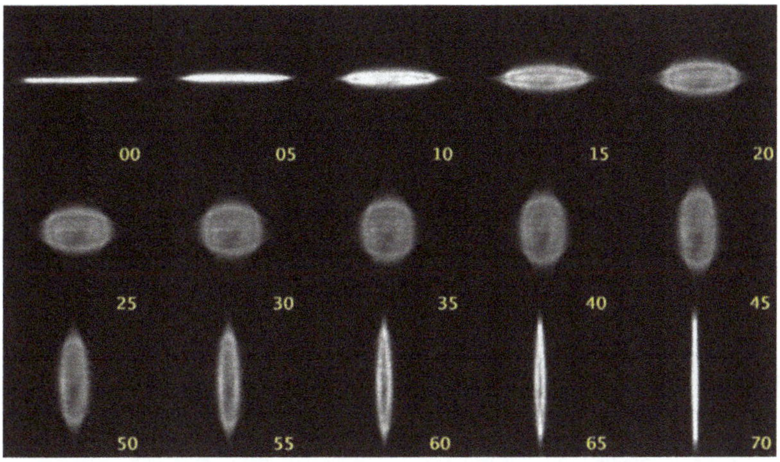

Figure 4.25. Anamorphic images of a fluorescent micron-sized particle traversed along the optical axis (the yellow numbers indicate the particle displacement in microns). Reprinted from [36] with permission of Springer.

images obtained with an APTV system, where it can be seen that the particle shape transitions from a horizontal elliptical profile to a vertical elliptical profile as the particle transitions through the depth of the imaged volume.

The uncertainty of the depth location is given as [36]

$$\Delta z = \frac{\delta d}{2(\mathrm{NA})M} = \frac{\delta s}{2(\mathrm{NA})^2}, \tag{4.32}$$

where δd is the uncertainty of the particle image diameter within the CCD plane, δs is the uncertainty in the wavefront sag measurement, M is the magnification, and NA is the numerical aperture. Furthermore, a compensation factor was developed to adjust for x and y particle location perspective errors. With calibration, the resulting depth position errors are 2.8 μm over a 500 μm depth range. Further studies by Cierpka *et al* [37] (see figure 4.26) show the sensitivity of astigmatism algorithms with a single camera at various noise levels using simulated data to the estimated to true axis ratio, ε_a (left), and to the subpixel position error (right), where NL is the ratio between the peak noise intensity levels and the peak image gray value. Here, particle image diameters were greater than three pixels. The plots show that the axis ratio and subpixel position error is about 0.5% and +/− 0.05 pixels for a 5% noise level, and that the subpixel position error increases to +/− 0.1 pixels for 20% noise levels.

Cierpka *et al* [37] proposed an intrinsic calibration method in order to extend the measurement volume's depth and to account for all image aberrations. In this work, they were able to image Poiseuille flow within a straight rectangular micro-channel with a cross-sectional areas 200 × 500 μm^2, and in-plane and out-of-plane velocity uncertainties of 0.9% and 3.7% of the centerline velocity, respectively. Seeding densities per image were 0.00006 N_{ppp}, and 2000 images were averaged to obtain

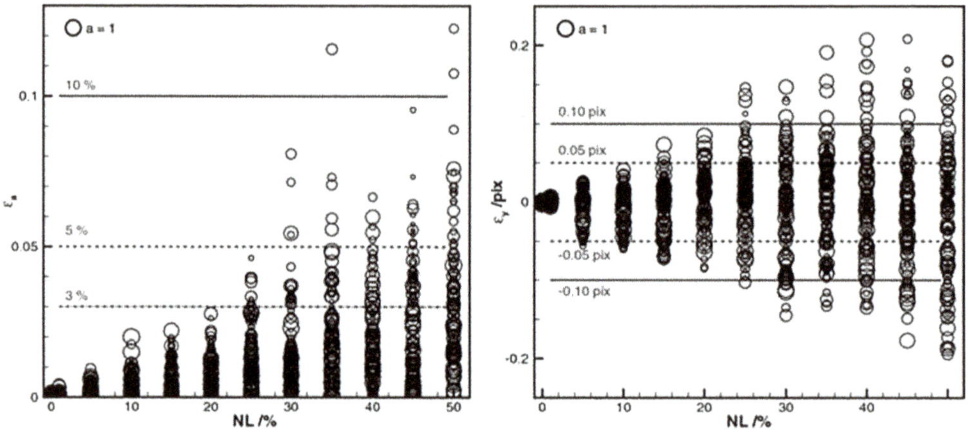

Figure 4.26. (left) Error in axis ratio and (right) in-plane displacement error with respect to noise level for single-camera astigmatism particle localization [37].

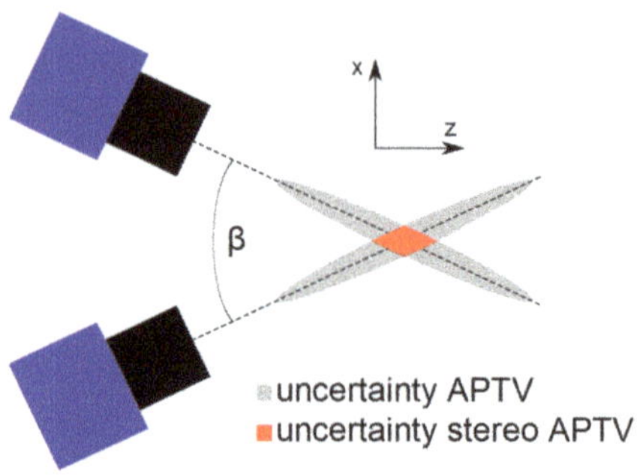

Figure 4.27. Qualitative uncertainty between astigmatism PTV techniques [41].

these results. In a later study, Rossi and Kähler [39], through an analytical study, identified three non-dimensional parameters, k_1, k_2, and k_3, that were most important in optimizing APTV performance. k_1 represents the ratio between the magnifications in the x and y directions, k_2 represents the ratio between the main focal planes' distance and the measurement volume depth, and k_3 represents the ratio of the diameter of the in-focus to the out-of-focus particles at the edges of the measurement volume. For optimized performance, they recommend that k_3 should be kept as low as possible ($k_3 < 0.15$), which states that the depth-of-focus should be small with respect to the measurement volume height; k_2 should be chosen depending on the calibration procedure used.

Fuchs *et al* [41] suggested adding a second astigmatism camera and using a stereoscopic PTV approach using triangulation when out-of-plane measurements of position and velocity are of the utmost importance. Figure 4.27 shows how the uncertainty is reduced with the addition of a second camera in that the stereoscopic uncertainty is the intersection of the uncertainty regions of each of the cameras. With 0.0005 N_{ppp} seeding densities, the statistical measurement error from stereoscopic astigmatism PTV was found to be in the range of 3–5 μm for x and y locations and 12–13 μm for z locations for an observation volume of $40 \times 40 \times 20$ mm^3, which is equal to an error in the z position estimate of 0.03% of d_z.

4.2.1.7 Particle diffraction pattern methods

Particle diffraction patterns can only be generated with clear interference fringes if the reflected or emitted light from the particle is coherent or partially coherent, and is then superposed constructively and destructively. This diffraction pattern is three-dimensional, extending both in depth through the optical axis, and radially about the depth axis. In the absence of aberrations, this pattern is symmetric, and asymmetric in the presence of aberrations [43]. The point spread function (PSF) is the analytical expression of this diffraction pattern, and its square corresponds to the intensity distribution seen within an image. Gibson and Lanni [44] derive a prediction for the square of the PSF that includes aberration, and based on Kirchhoff's scalar diffraction integral formula, as

$$I(x_d, y_d, \Delta z) = \left| C \int_0^1 J_0 \left[k \frac{\mathrm{NA}}{\sqrt{M^2 - \mathrm{NA}^2}} \rho \sqrt{x_d^2 + y_d^2} \right] \exp\left[jW(\Delta z, \rho) \right] \rho d\rho \right|^2. \quad (4.33)$$

Here, NA is the numerical aperture, M is the magnification, (x_d, y_d) is the non-dimensional lateral particle locations, Δz is the depth location from the focal plane, k is the wave number, ρ is the normalized radius in the exit pupil of the objective, and $W(\Delta z, \rho)$ is the phase aberration function. The phase aberration function is a product of the wave number and the optical path difference (OPD), which is an important element in addressing optical aberrations. Detailed derivations and expressions for the OPD can be found in Gibson and Lanni [44] and Park [45].

An example of particle diffraction patterns is shown in figure 4.28. Speidel *et al* [46] showed that particle diffraction patterns can be used to map 3D flow fields. Using a single nanoparticle within an epifluorescent microscopic imaging system,

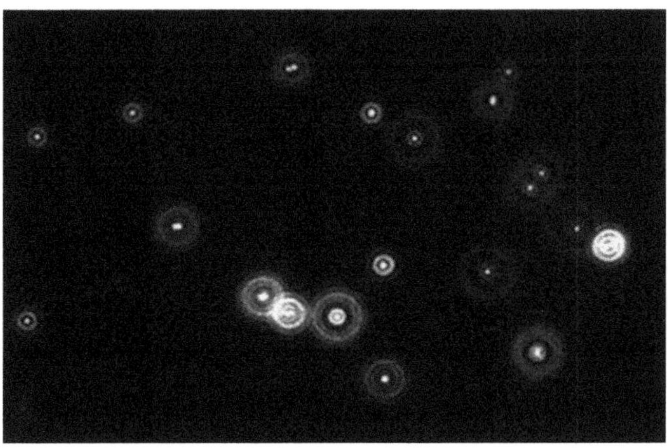

Figure 4.28. Flow image at 0.225 s [51].

they demonstrated the concept by moving a 216 nm fluorescent microbead away from the focal plane, showing that the particle first blurs, then multiple rings appear, where the outermost ring is the brightest. In this regard, they called this approach the optical serial sectioning microscopy (OSSM). They then calibrated the outermost ring to the depth location. Wu *et al* [47] demonstrated this approach also works for micrometer scale particles, as well as for imaging multiple microbeads. In addition, through simulation, they showed that the outermost ring is dominated by spherical aberrations.

Park and Kihm [48] calibrated the outermost diffraction ring to the particle's depth location with respect to the focal plane. An image processing algorithm was developed to identify the particles' center locations as well as the diameter of the outermost diffraction rings. An analytical model was formulated to create a calibration curve to calibrate the outermost diffraction rings to the depth location. Figure 4.29 shows the analytical evaluations of the diffraction patterns at different depth locations in (b) and (c) and the depth the outermost diffraction ring size

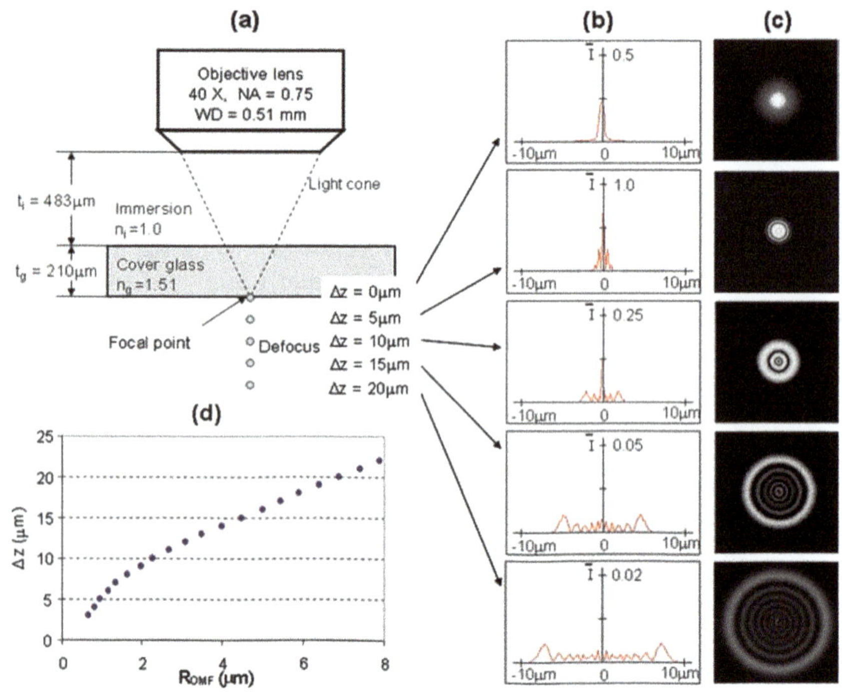

Figure 4.29. (a) A deconvolution microscopy experimental set-up for 3D-PTV, with critical physical parameters. (b) PSF of diffracted particle image at each indicated defocus distance. (c) Graphical diffraction images derived from the PSF. (d) The relationship between defocus distance (Δz) and out-most fringe radius (R_{OMF}) [48].

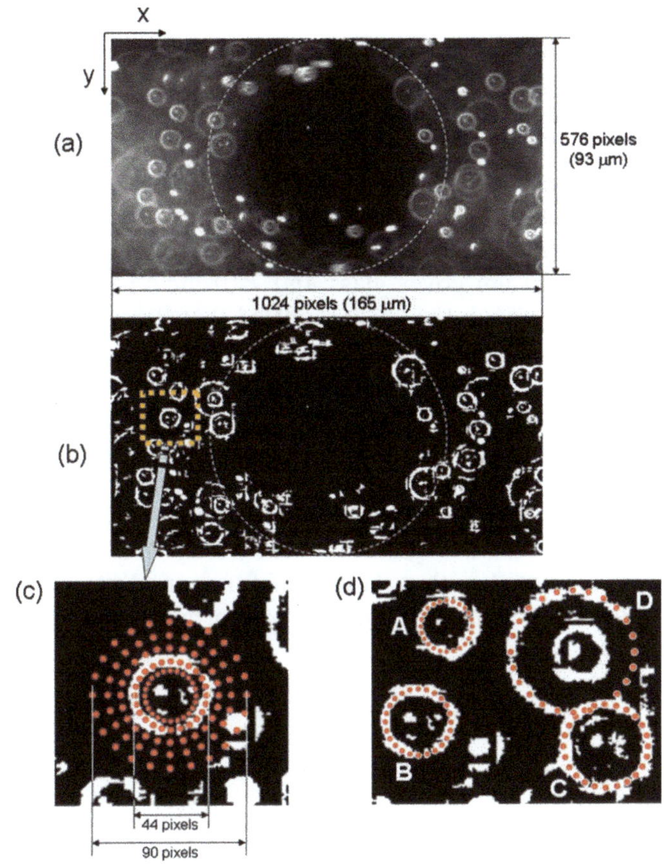

Figure 4.30. (a) A raw diffraction image of flow around a micro-sphere, captured with an intensity resolution of 8-bit and a shutter speed of 30 ms. (b) Image processing was used convert the image from grayscale to binary. (c) Three-dimensional particle positions were identified using axisymmetric 24-point data. (d) A few identified three-dimensional particle positions: (A) *x*-, *y*-, R_{OMF}-pixel numbers: 158, 178, 16; (B) 140, 245, 22; (C) 240, 200, 40; and (D) 264, 261, 29. Reprinted from [48] with permission of Springer.

relation in (d). Figure 4.30 illustrates the image processing algorithm they developed for image flow around a sphere. Figure 4.30(a) shows a raw CCD image on which this algorithm is applied. Figure 4.30(b) illustrates the resulting image after a multi-step binarization process. Figure 4.30(c) illustrates the circle-fitting process. Conceptually, a sufficiently large circle centered around each pixel is started, and is then incrementally shrunk, whereupon each iteration, the brightness of the circumference is examined. Then the outermost rings of each particle are identified, and its center is assigned as the *x*–*y* location of the particle, and its depth is determined from the calibration results obtained as shown in figure 4.29(d). Figure 4.30(d) shows examples of particle positions that have been identified using this procedure.

Luo *et al* [49] also used the Gibson and Lanni [44] Kirchhoff's scalar diffraction model but with an OPD based on Haebelé [50] to predict the diffraction pattern details through a glass cover, which introduced a mis-matched refractive index, within an air immersed microscope objective. Rather than using only the outermost diffraction ring, they used a combination of first dark and first bright fringe diameters. In a follow-up study, Luo *et al* [51] studied the effects of cover glass thicknesses and the use of partially coherent light emanating from fluorescent particles on the diffraction pattern sizes and their effect on particle depth identification. For partially coherent light, they found that the wavelength distribution of the fluorescence significantly affected the particle diffraction pattern at large depth locations. Here, they were able to show the predictions worked well for small depths (<30 μm for 1100 μm cover glass and <10 μm for 160 μm coverslip) for monochromatic light, although they also reported that the scalar diffraction model overestimated the diffraction pattern sizes. In addition, for the smaller depths, the central interference fringes were used to estimate particle depth locations, while for larger depths, the outermost fringes were used to make this estimation. In addition, they found that smaller thicknesses of the cover glass produce larger changes in the diffraction pattern through depth, resulting in shorter depths where particles can be tracked. In further work, Luo and Sun [52] noted that although diffraction patterns change with depth, their central regions are nearly Airy disk intensity distributions, which can be approximated with a Gaussian function. With this in mind, they developed an algorithm that iteratively cross-correlated a Gaussian intensity distribution template

$$g(r) = g_0 \exp\left(\frac{-r^2}{2\Delta^2}\right), \tag{4.34}$$

where g_0 is the maximum intensity at the template's center, r is the radial distance from the center, and Δ is the Gaussian peak width, through their image using

$$c(i, j) = \sum_{m=1}^{m=M} \sum_{n=1}^{n=N} g(m, n) \cdot f(i + m,\ j + n), \tag{4.35}$$

where c is the cross-correlation matrix, ($\cdot$) denotes a cross-correlation operation, and (M, N) denote the size of the Gaussian template, to identify diffraction centers. Once particle centers were identified, their earlier approaches were used to identify diffraction patterns, which were then cross-correlated with a calculated Gibson–Lanni distribution, using a normalized cross-correlation algorithm, to identify the particles' depth locations. Using this approach, the particles' centers were found with uncertainties within 0.02–0.04 μm (0.1–0.2 pixels). For depth locations greater than 13 μm, the diffraction patterns contained at least three fringe rings, and using a 1.1 mm cover glass slide the accuracy was recorded to be better than 0.1 μm, or an order of magnitude larger than the particles' center accuracies.

Peterson *et al* [53] proposed capturing calibration images of a particle at various known depth locations to create a calibration curve. From these images, the particle's center was determined, and the outermost ring was identified, which was used to calibrate against the known depth locations. To generate the calibration curve, the particle diameters, 0.3, 1, and 3 μm were tested with four different microscopic lenses with magnifications of 20×, 40×, and 50× and effective numerical apertures of 0.6, 0.8, and 0.73, respectively. A quadratic was used to fit the data with determination coefficients that were better than 99.8%. Uncertainties in particle center locations averaged to 0.6 pixels; uncertainties in depth are directly related to determination of the outermost ring, which was 0.08 pixels for the larger rings, and 3.4 pixels for the smaller rings.

4.2.2 Tomographic PTV

Both the purpose and the set-up of a tomographic PTV (tomo-PTV) system are similar to that of a photogrammetric PTV system. Figure 4.31 shows an example set-up with four cameras. A tomographic PIV algorithm implements the steps outlined in figure 4.31 for image reconstruction and cross-correlation analysis. The key step in this algorithm that translates exactly to tomo-PTV is the tomographic

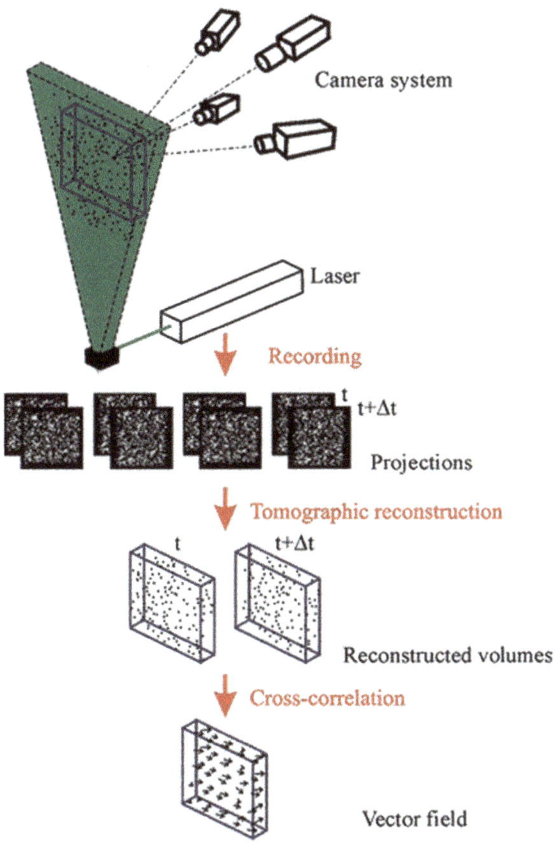

Figure 4.31. Four-camera tomographic PTV set-up. Reprinted from [54] with permission of Springer.

reconstruction. This is done using CCD pixel intensity data before any particle identification. Unlike other methods, the particle identification step is done in the three-dimensional space as a single step. In addition, a particle mask correlation algorithm (section 3.2.3) has been extended into 3D for use in the voxel space for tomo-PTV [55]. On the other hand, voxel clusters can also be identified and the centroid can be estimated from three 1D Gaussian fits (section 3.1.3) [57].

The tomographic reconstruction algorithms for PIV and PTV are generally based on the multiplicative algebraic reconstruction technique (MART) [55]. The MART technique is based on image particle intensities captured by each camera and the correlation with the actual illuminated particle intensity. This relationship is given in equation (4.32), where E_i is the intensity of pixel i and $I_{x,y,z}$ is the light intensity at position x, y, z,

$$E_i = \int I_{x,\,y,\,z} dl. \tag{4.36}$$

In order to discretize and relate the x, y, z coordinates to the camera's pixels, a discretized 3D domain is created. These discrete elements are called voxels, and make up the measurement volume. In this discretized form, equation (4.36) is modified as shown below in equation (4.37). A weighting factor, w_{ij} is introduced to relate each pixel intensity to each voxel intensity. These weighting factors rely heavily on an accurate calibration procedure in three dimensions. The 2D grid of markers is positioned at various depths within the measurement volume to obtain the proper mapping coefficients (figure 4.32):

$$E_i = \sum\nolimits_j w_{ij} I_j. \tag{4.37}$$

Iterations of the MART algorithm are calculated using this concept. Initially, the voxel field is set to a uniform, nonzero value. Values of I_j are then iteratively updated using equation (4.38),

$$I_j^{k+1} = I_j^k \left(\frac{E_i}{\sum_j w_{ij} I_j^k} \right)^{\mu w_{ij}}. \tag{4.38}$$

An alternative algorithm for tomographic reconstruction is the multiple line-of-sight algorithm (MLOS), which generates voxel intensities from a single pass. MLOS is less accurate than MART, although using it as a first guess to allow MART to only operate on nonzero voxels can decrease the processing time by a factor of 5–10 [18]. Simultaneous multiplicative algebraic reconstruction (SMART) is another technique that closely resembles MART, and can further decrease computation time by performing some simultaneous calculations. The MLOS-SMART algorithm was shown to maintain the same accuracy as MART while decreasing computation time by a factor of 15 [56].

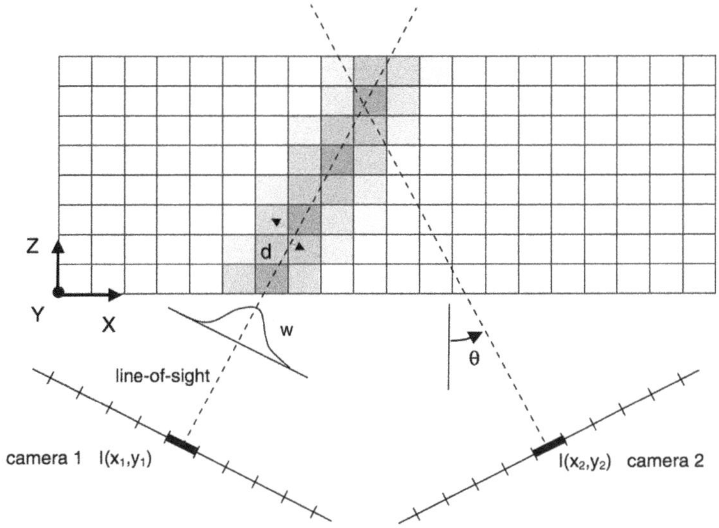

Figure 4.32. 2D representation of pixel and voxel interaction in tomo-PTV. Reprinted from [54] with permission of Springer.

Once converged, the tomographic reconstruction will yield a 3D voxel space containing particle images. Using this voxel space, the particle identification algorithms explained in chapter 3 can be used. Both the image segmentation and Gaussian centroid estimation steps can be extended for use in three dimensions.

Kitzhofer and Brücker [57] suggested that for small interrogation volumes, tomographic reconstruction can be simplified by using telecentric imaging. Telecentric lenses have constant magnification and parallel projection, making the relationship between pixels and voxels simpler, as shown in figure 4.33. This simplification reduces computation time and performs similarly to MART reconstruction for particle seeding densities up to 0.1 N_{ppp}, although for a volume with depth equal to its height and width, the authors suggested using the algorithm for tracking with 0.01 N_{ppp}. The use of telecentric imaging limits the imaging volume dimensions to those of the lenses used. In the case of Kitzhofer and Brücker, the lenses limited the volume to $90 \times 90 \times 90\ \text{mm}^3$.

Tomographic reconstruction can be ambiguous, leading to what are called ghost particles in the reconstruction volume. High seeding particle density and too few cameras increase the probability of ghost particles appearing within the reconstruction. Elsinga *et al* [54] defined a quality factor Q

$$Q = \frac{\sum_{X,Y,Z} E_1(X,\ Y,\ Z) E_0(X,Y,Z)}{\sqrt{\sum_{X,Y,Z} E_1^2(X,\ Y,\ Z) \sum_{X,\ Y,\ Z} E_0^2(X,\ Y,\ Z)}}, \tag{4.39}$$

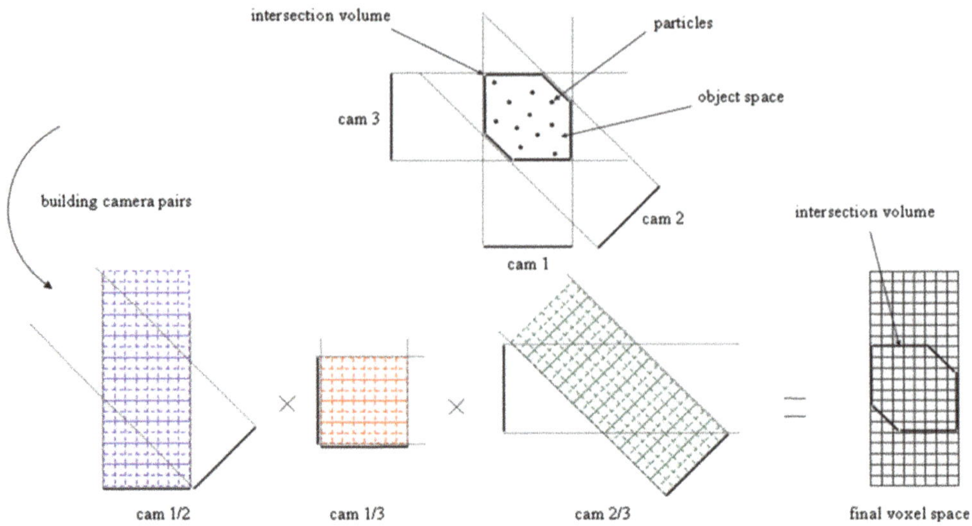

Figure 4.33. Volume definition with pixel interaction for telecentric imaging [57].

to quantify reconstruction volume accuracy, which is a normalized correlation coefficient between the reconstructed image, $E_1(X, Y, Z)$ and the exact distribution of light intensity, $E_0(X, Y, Z)$. The exact distribution is generated by a Gaussian intensity distribution of three voxel diameter centered at each known particle location. The quality factor as it relates to the number of cameras, particle density, calibration error, and viewing angle (angle between the outer cameras and the z-axis) can be seen in figure 4.34. Three cameras at $\theta = \{-20, 0, 20\}$ degrees, 0.05 N_{ppp}, five reconstruction iterations, and no calibration errors or image noise. A two-camera system is largely insufficient, and adding cameras increases Q rapidly as it approaches unity with five cameras. The optimal angle between cameras is 30 degrees for the three-camera set-up, as larger angles increase the intercepted length of the line-of-sight between cameras, causing a larger number of ghost particles to appear. Increasing the particle seeding density also increases the number of ghost particles, although increasing the number of cameras allows for higher seeding densities. Experimental calibration error also reduces Q. Reconstruction quality greater than 0.75 was achievable for errors below 0.4 pixel.

4.2.3 Synthetic aperture PTV

The synthetic aperture method, first introduced for PIV by Belden *et al* [58] and extended to Lagrangian particle tracking by Bajpayee and Techet [59], utilizes five or more cameras (nine were used in this study) to simulate a lens with an arbitrarily large aperture. By backprojecting particle images from each camera to a specific depth and merging the images, particles present at the chosen depth will be sharply focused. Particles located at a different depth will be blurred and dim. Performing this refocusing at depths throughout the volume of interest gives a 3D reconstruction of particle locations.

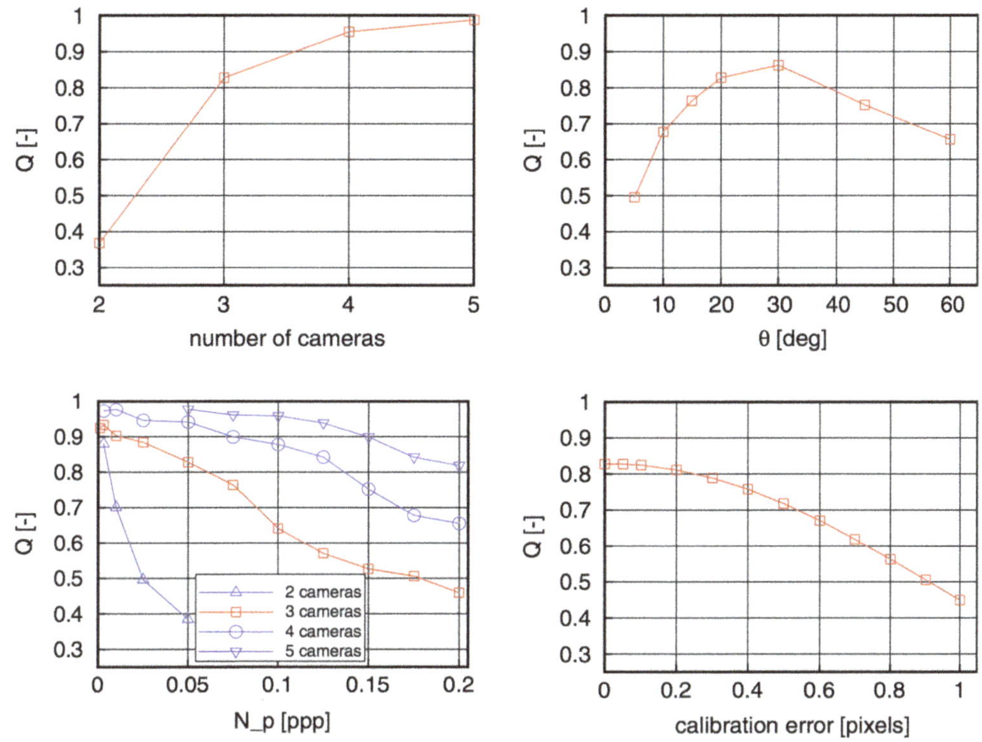

Figure 4.34. Tomographic reconstruction quality compared to exact intensity distribution as a function of set-up and calibration. Reprinted from [54] with permission of Springer.

A reference plane within the observation volume is defined by the location of the planar calibration grid. A homography transform from the ith camera frame to the reference plane is defined based on a calibration

$$\begin{Bmatrix} bx' \\ by' \\ b \end{Bmatrix}_i = \begin{bmatrix} h_{11} & h_{12} & h_{13} \\ h_{21} & h_{22} & h_{23} \\ h_{31} & h_{32} & h_{33} \end{bmatrix} \begin{Bmatrix} x \\ y \\ 1 \end{Bmatrix}_i, \tag{4.40}$$

where b is a constant, x and x' are the image and reference plane coordinates, and h_{jk} are coefficients obtained from camera calibration. The reference plane particle images are then refocused to new focal planes in the observation volume using

$$\begin{Bmatrix} x'' \\ y'' \\ 1 \end{Bmatrix}_i = \begin{bmatrix} 1 & 0 & \mu_k \Delta X_{Ci} \\ 0 & 1 & \mu_k \Delta Y_{Ci} \\ 0 & 0 & 1 \end{bmatrix} \begin{Bmatrix} x \\ y \\ 1 \end{Bmatrix}_i, \tag{4.41}$$

where x'' is the new focal plane coordinate, ΔX_{Ci} and ΔY_{Ci} are the relative camera locations of the ith camera, and μ_k is a constant that determines the location of the kth focal plane. The camera locations can be determined using a calibration given by Vaish *et al* [60], although a calibration image taken at each focal plane can be used

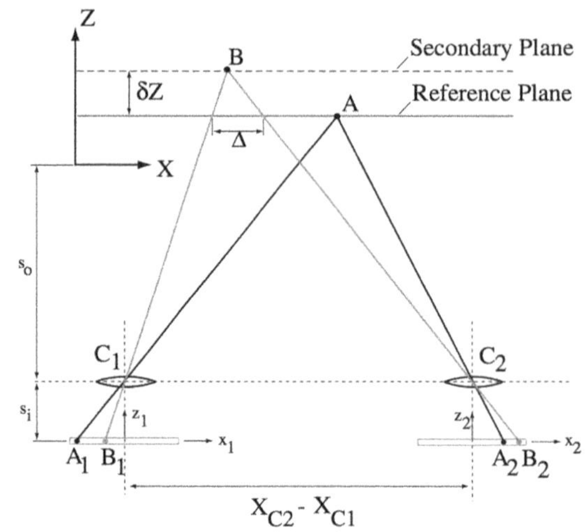

Figure 4.35. Schematic showing the reference plane and focal plane as well as illustrating the concept of parallax [58].

to determine the h and μ coefficients for each camera (see section 4.3.3). The refocused images from all the cameras are then averaged to create a single image for each focal plane.

Figure 4.35 shows how two cameras capture the images of a point in a secondary plane, B, and in a reference plane, A. When the two images are remapped to the reference plane, the images of particle A will align and appear as a bright coherent point while the images of particle B will appear as two separate, dim points. This is due to the parallax between cameras and allows in-plane points to be differentiated from out-of-plane ones.

Modifications to the refocusing transformation in the presence of refractive interfaces were introduced by Belden [61]. The computational cost of 3D reconstruction is high when each pixel is backprojected individually through refractive interfaces, therefore Bajpayee and Techet [62] suggested a homography transform to approximate the distortion when the interfaces are planar. This homography transform, H, is calculated by projecting the four corner points of each camera through the refractive interfaces:

$$P_j^{'Z} = HR_j^{'Z}. \tag{4.42}$$

$P_j^{'Z}$ are backprojected corner points from the jth camera onto the z plane and $R_j^{'Z}$ is the projection of these points onto the camera. This transformation is shown in figure 4.36. The homography fit approximates refraction through planar surfaces only.

In the SAPTV algorithm used with synthetic data, the average localization error was found to be 0.2177 mm for a reference depth of 5 mm. In an experiment with a

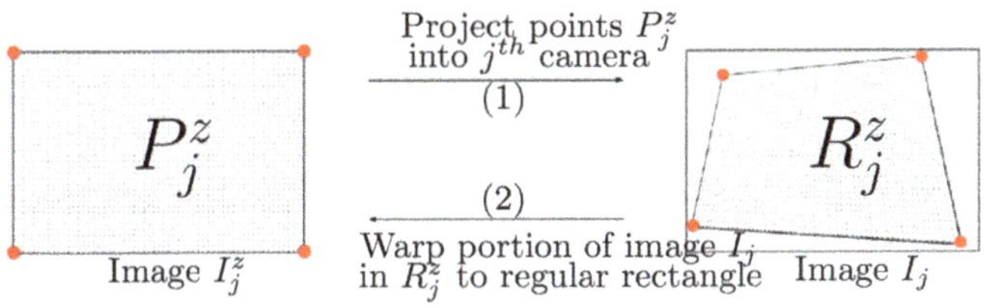

Figure 4.36. Projection and backprojection using the homography fit method for synthetic aperture refocusing. Reprinted from [54] with permission of Springer.

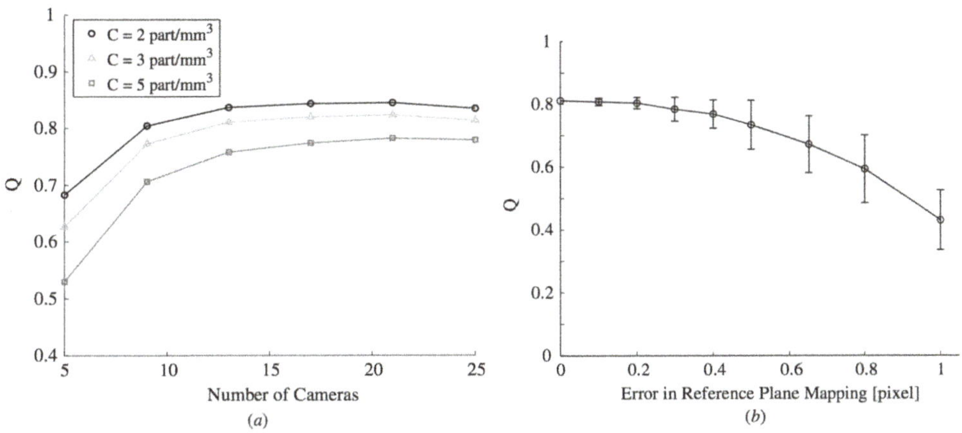

Figure 4.37. Reconstruction quality, Q, as a function of the number of cameras (a) and reference plane mapping error (b) in a $50 \times 50 \times 10$ mm^3 volume. C is the particle seeding density [58].

vortex ring, between 420 and 480 particles were detected in each frame, and 366 particles were visible for 1 s at 30 Hz [59].

Belden *et al* [58] used the synthetic aperture method to reconstruct 0.125 N_{ppp} in a simulation and 0.015 N_{ppp} in an experiment. They also showed that the reconstruction quality, defined by Elsinga [55], depends on the number of cameras used and the mapping error onto the reference plane. Figure 4.37 shows that the synthetic aperture method performs best with 13 cameras and has less than 0.2 pixels mapping error in the reference plane. When compared to tomo-PTV, the homography fit synthetic aperture reconstruction increased the computational efficiency by 270 times when used for a $512 \times 512 \times 128$ voxel volume and 0.1 N_{ppp} [62].

4.2.4 Plenoptic imaging

A plenoptics camera uses a microlens array to image a 3D volume in a manner that captures both the position and angle of the light rays. In doing so, the complete distribution of the light rays within the volume can be functionally expressed, which is denoted as the light field. Using these data, the image can be computationally refocused to reconstruct the particle field within imaged volume in three dimensions. The set-up of the microlens array is shown in figure 4.38. On the left, a conventional imaging set-up is shown, where the light rays emanating from a spatial coordinate

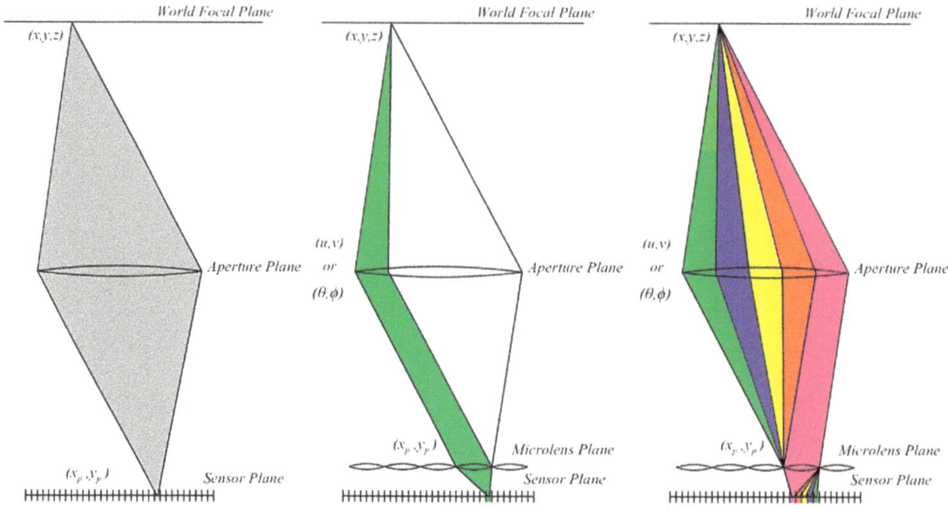

Figure 4.38. Schematic showing the application of micro-lenses for capturing the angle of incoming light in a plenoptic camera [67].

are collected by the lens and focuses onto the sensor plane. Note that this set-up does not provide any information about each of the rays' position and angle. With the proper placement of the microlens array, rays within the green region are refocused onto the sensor plane, as shown in the center figure. On the right, when now looking at different rays emanating from the spatial coordinate, it can be seen that the microlens array refocuses them onto different regions of the sensor plane. In this manner, each pixel is able to view the different angles of the light rays, thereby capturing the total light field content of the imaged volume.

The refocusing process [64–66] can be demonstrated using figure 4.39. Given an acquired image at the microlens plane, and its geometric relation to the aperture plane, the virtual light field L' at the virtual x' plane located at s_i' can be reconstructed from the original light field L,

$$L'(x', y', u, v) = L\left(u\left(1 - \frac{1}{\alpha}\right) + \frac{x'}{\alpha}, v\left(1 - \frac{1}{\alpha}\right) + \frac{y'}{\alpha}, u, v\right), \tag{4.43}$$

where (x', y') is the virtual image sensor, and $\alpha = s_i'/s_i$. The refocused image at the virtual sensor planes can be calculated by integrating the above equation, resulting in

$$I(x', y') = \int\int L\left(u\left(1 - \frac{1}{\alpha}\right) + \frac{x'}{\alpha}, v\left(1 - \frac{1}{\alpha}\right) + \frac{y'}{\alpha}, u, v\right) dudv. \tag{4.44}$$

Tomographic methods for volume reconstruction of plenoptic imaging data have been used to eliminate out-of-plane images [67–69]. These algorithms use the MART algorithm described in section 4.2.2. The 3D reconstruction is discretized into voxels, just as in tomo-PTV; however, the size of these voxels is chosen to be similar to that of a microlens rather than that of a pixel. The technique used in the

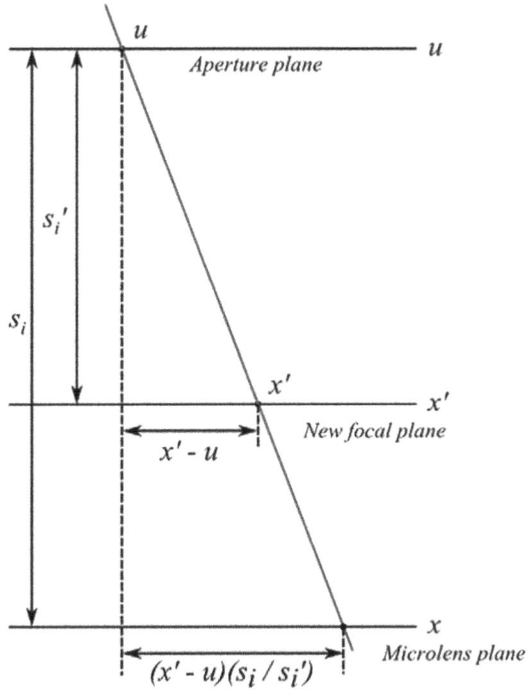

Figure 4.39. Illustration of interpolation for refocusing using two-plane parameterization [68]. Adapted from Ng [63].

refocusing algorithm is used here to find a weighting function, which defines the relationship between the image space and the voxel space. In order to define this weighting function, the plenoptic camera is mapped to each plane of voxel elements and a linear interpolation is performed to determine the intersection of each pixel with each voxel. The weighting coefficients are then normalized such that the sum of weights for each voxel is unity. The technique for determining this weighting function is described in detail by Farhinger *et al* [67].

The MART plenoptic 3D reconstruction was tested using synthetic data in a $37.8 \times 25.0 \times 25.0$ mm^3 volume, which was discretized into $300 \times 200 \times 200$ voxels^3. The errors in the algorithm were less than one voxel in-plane and three voxels in depth. Particle densities between 0.0001 and 0.01 N_{ppp} were tested. The reconstruction quality, Q, was above 0.9 for densities up to 0.0032 N_{ppp}, and dropped to 0.23 at 0.01 N_{ppp}.

Farhinger and Thurow [69] later suggested using a filtered refocusing algorithm, which calculates refocused intensities throughout the observation volume then filters based on the fraction of projections that exceed an SNR threshold. Each pixel that contributes to a voxel's intensity is then checked against this threshold. The ratio of

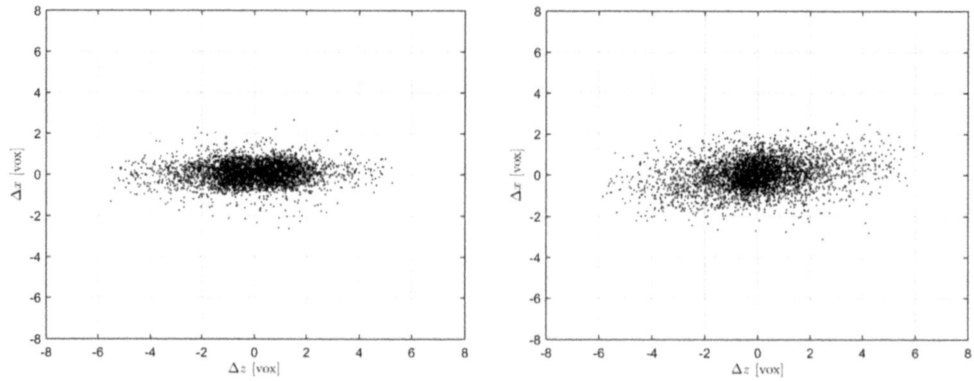

Figure 4.40. Errors in plenoptic imaging reconstructions using filtered refocusing (left) and MART (right) [69].

these valid projections and the total projections is then compared to a second user-selected threshold. If the threshold is exceeded, the reconstructed voxel intensity is used as normal; if the threshold is not met, the voxel intensity is set to zero.

Figure 4.40 shows a comparison of the errors of this filtered refocusing algorithm and the MART based algorithm. The standard deviation of in-plane position errors were 0.519 voxels and 0.746 voxels for filtered refocusing and MART, respectively. The errors in depth position estimates were 1.571 and 1.753 voxels.

Anglin *et al* [70] developed a deconvolution method for volume reconstruction of plenoptic data in order to reduce the computational cost of volume reconstruction. The refocusing algorithm is inherently shift-variant, thus for the purpose of deconvolution, a shift-invariant form was developed. A further benefit of this approach is that the deconvolution algorithm was able to reconstruct a volume in seconds, compared to the hours required for MART based reconstruction.

Figure 4.41 demonstrates the differences between the MART reconstruction and the deconvolution algorithm. Deconvolution introduces image blur in the depth direction, causing roughly double the localization error compared to MART. The MART algorithm, on the other hand, failed to recover all the points in the reconstruction.

Chen and Sick [71] performed an experimental analysis of a plenoptic 2.0 camera, which is designed such that the microlens array is placed farther than one focal length away from the camera sensor. In doing this, the angular information is compromised for spatial resolution. The camera used was a 29-megapixel color plenoptic camera. The effective resolution of this camera was 7 megapixels. In this application of Plenoptic PTV, the particle image positions were identified in each microlens using a particle mask correlation algorithm (section 3.2.3). Each particle had 14 corresponding images on different micro-lenses, allowing for robust determination of 3D position with only one camera.

In the context of the increased effective resolution of plenoptic 2.0 cameras, Chen and Sick showed that seeding densities should be kept below 0.25 particles per microlens in order to achieve good tracking results. They showed that error in displacement calculations was minimal even at these high seeding densities. The

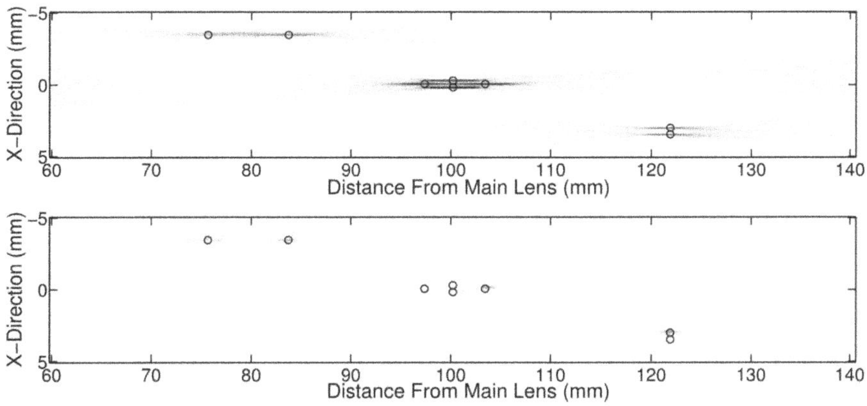

Figure 4.41. Reconstructed 3D volume using deconvolution (top) and MART (bottom) algorithms. 'o' indicates point locations and inverted grayscale is used [70].

impact of out-of-plane particle motion on particle displacement was tested; it was found that out-of-plane displacements greater than 2% of the measurement depth were overestimated by 20%, and an out-of-plane displacement of 16% of the measurement depth caused a 25% overestimate of in-plane displacement. It was suggested that the middle 40% (20 mm) of the examined depth (50 mm) gave accurate 3D reconstruction results for displacements less than 8 pixels (0.4 mm).

Generally, plenoptic cameras are effective for volume reconstruction when optical access is limited, as only one camera is used with one viewing angle. Accurate reconstructions are achievable using either computationally expensive MART algorithms, or faster deconvolution. The achievable seeding densities of these algorithms is limited due to the effective reduction in spatial resolution.

4.2.5 Holographic PTV

This section discusses the basics of digital PTV holography as well as its state-of-the-art techniques for 3D reconstruction. For a more thorough discussion, the reader is referred to one of several review papers [72–74].

Holographic imaging is a technique for capturing 3D information by using a film or digital camera to record a reference light source superimposed with the light scattered by an object. The illumination must be a coherent laser light source so that the object and reference interfere constructively and destructively. The holographic recording, I_h, captures the interference between a complex reference wave, $\tilde{R}$, and an object wave, $\tilde{O}$,

$$I_h(x_h, y_h) = (\tilde{R} + \tilde{O})(\tilde{R}^* + \tilde{O}^*) = RR^* + \tilde{R}\tilde{O}^* + \tilde{R}^*\tilde{O} + \tilde{O}\tilde{O}^*, \tag{4.45}$$

where the asterisk indicates the complex conjugate and (x_h, y_h) represent the hologram image coordinates. The first term of this equation represents the mean intensity of the reference beam. The second and third terms generate the interference patterns used to determine depth. The final term represents the speckle pattern effect [72]. The interference patterns have a fringe spacing given by

$$d_f = \frac{\lambda}{2 \sin \alpha}, \tag{4.46}$$

where λ is the wavelength of the coherent light source and α is the half-angle between the directions of light propagation of the two interfering waves. The fringe spacing is a maximum when the object wave and reference wave are aligned. Holographic imaging for PTV is done using either an in-line or an off-axis set-up, referring to the alignment of the reference and object light sources. An in-line system uses a single laser for object illumination and as a reference wave, whereas off-axis systems will use two separate lasers. The low resolution of digital cameras limits experiments to in-line holography, as the fringe spacing in off-axis holography requires high resolution film [75]

The greatest limitation of holographic imaging is the low spatial resolution in the depth direction. This is a result of the finite aperture and it causes reconstructed particle images to be elongated in the depth direction. In-line holographic systems typically elongate particles to be in the range of several millimeters up to tens of millimeters while off-axis systems only elongate particles to 0.2–0.5 mm [75].

Typical in-line and off-axis holographic systems are depicted in figures 4.42 and 4.43, respectively. Note that the in-line set-up is simple and does not require holographic film. In-line holography also benefits from forward scattering, requiring only 0.1% of the laser power of standard 3D-PTV methods [76]. Off-axis systems have better depth resolution than in-line systems and also do not suffer from the speckle pattern effect. Relative ease of use has caused digital in-line systems to be the most common for PTV applications [75].

Digital in-line holography is particularly well-suited for microscale experiments, as the depth-of-field can be extended to almost 1000 times that of a conventional microscope. The effect of particle elongation is also reduced to only 2–10 times the particle diameter, decreasing with increasing magnification [72]. The only required modification to the experimental set-up in figure 4.42 is the addition of a microscope objective lens, as shown in figure 4.44.

A PTV hologram captured using in-line recording is shown in figure 4.45(a). The interference patterns that are generated by particles appear as fringe patterns with spacing that varies depending on a particle's depth position. The fringe patterns

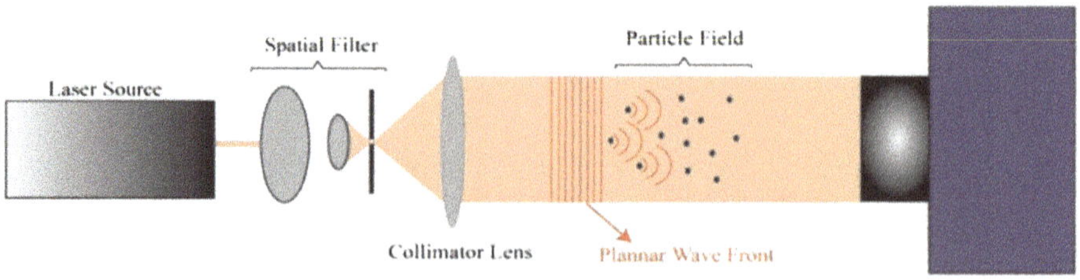

Figure 4.42. Typical lens-less digital in-line holographic PTV set-up [77].

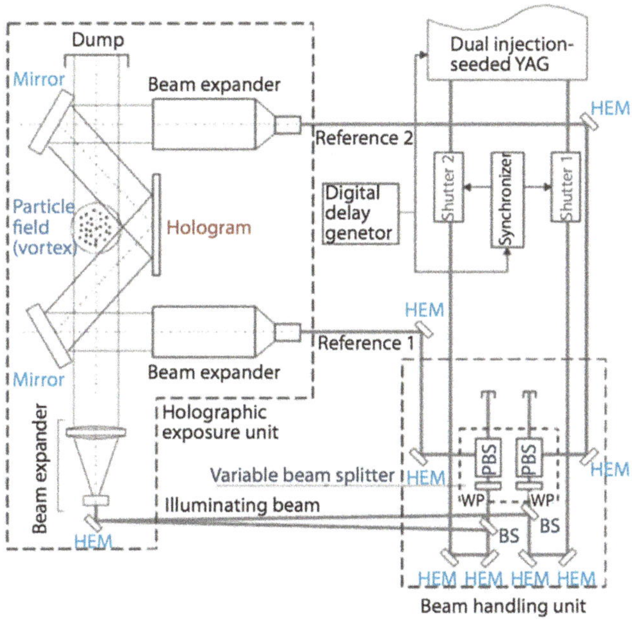

Figure 4.43. Double-exposure off-axis holographic PTV set-up using two separate reference beams [78].

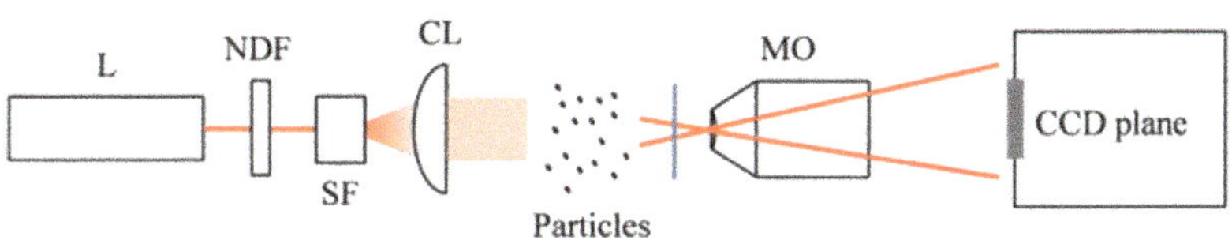

Figure 4.44. Microscopic digital in-line holographic PTV set-up. Reprinted from [72] with permission of Springer.

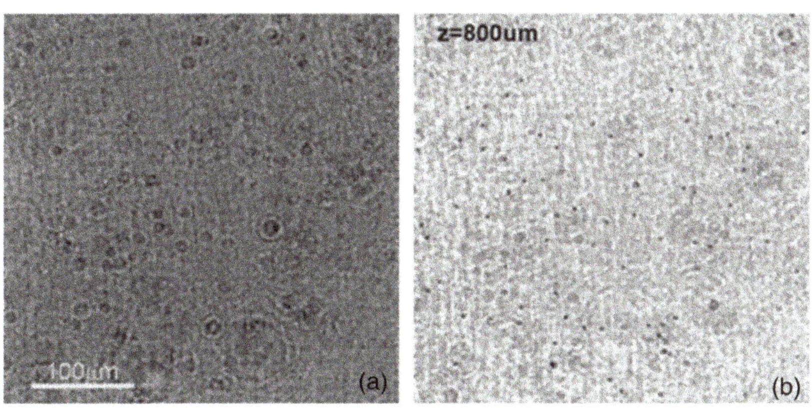

Figure 4.45. Part of a raw holographic recording (a) and a reconstruction of the plane located 800 μm from the hologram plane (b). The in-focus particles appear as dark spots on the bright background. Reprinted from [73].

overlap and interact with those of surrounding particles, causing a speckle pattern effect. This is the cause of the granular appearance of holograms. Volume reconstruction algorithms refocus the raw hologram to different depth planes. One such reconstructed plane is shown in figure 4.45(b). The fringe patterns of in-plane particles become dark, focused points.

Digital reconstruction is performed by approximating the optical field as a diffraction pattern that propagates from the hologram plane. The 3D field reconstruction is generated by

$$\tilde{U}_r(x, y, z) = \iint_{x_h, y_h} \tilde{U}_r(\xi, \eta, z = 0)\left[-\frac{\partial G}{\partial n}(x - \xi, y - \eta, z)\right]d\xi d\eta, \tag{4.47}$$

where $\frac{\partial G}{\partial n}$ represents the normal derivative of the Green's function of the wave equation propagating in a homogeneous medium evaluated at the aperture plane, $z = 0$. This term represents the diffraction of the tracer particles and is used for the 2D convolution integral with the hologram. The most common approximation of this diffraction term is the Rayleigh–Sommerfeld formula

$$-\frac{\partial G}{\partial n}(x, y; z) = \frac{1}{\lambda}\frac{\exp(-jk\sqrt{x^2 + y^2 + z^2})}{\sqrt{x^2 + y^2 + z^2}}\cos\theta, \tag{4.48}$$

where $\cos\theta = z/\sqrt{x^2 + y^2 + z^2}$ and is generally assumed to be roughly unity for simplicity, λ is the laser wavelength, j is the unit imaginary number, and $k = 2\pi/\lambda$. This equation can be simplified using binomial expansion and a paraxial approximation, which gives the Kirchhoff–Fresnel expression

$$-\frac{\partial G}{\partial n}(x, y; z) = \frac{\exp(-jkz)}{j\lambda z}\exp\left[j\frac{k}{2z}(x^2 + y^2)\right]. \tag{4.49}$$

The 3D reconstruction is most commonly simplified by analysis in the frequency domain, such that

$$\tilde{U}_r(x, y, z) = \mathcal{F}^{-1}[\mathcal{F}\{\tilde{U}_r(\xi, \eta, z = 0)\}\mathcal{F}\{\partial G/\partial n\}], \tag{4.50}$$

where $\mathcal{F}$ represents the Fourier transform. The reconstruction is performed for each z plane throughout the imaged volume. The image at a certain point is given by the irradiance field, $\tilde{U}_r\tilde{U}_r^*(x, y, z)$[72].

After volumetric reconstruction, voxel elements contain images of particles, which are elongated in the depth direction. This effect is commonly called the depth-of-field problem and is caused by the effective aperture angle, which can be limited by camera resolution, the angle of light diffraction, or the camera size. The elongation, τ, of a particle image is

$$\tau = \frac{\lambda}{\theta^2}, \tag{4.51}$$

where λ is the wavelength of the coherent light source and θ is the aperture half-angle. When using no magnification, the large pixel size of a digital camera can cause uncertainty in the location of interference fringes. The effective aperture angle of a resolution-limited hologram is

$$\theta = \tan^{-1}\left(\frac{\lambda}{2\Delta x}\right), \tag{4.52}$$

where Δx is the sensor pixel size. Magnification causes this resolution-limit to decrease, causing a hologram to be limited by light diffraction. Assuming Mie scattering, the effective aperture angle is given by

$$\theta = \frac{\lambda}{d}, \tag{4.53}$$

where d is the particle diameter. This aperture angle is unaffected by the magnification; thus, magnification should be chosen such that a hologram is diffraction-limited to minimize the depth-of-field. If the magnification is further increased, a hologram can become size-limited, meaning that the field-of-view of a camera limits the depth resolution. The size-limited aperture angle is

$$\theta = \tan^{-1}\left(\frac{N\Delta x}{2MZ}\right), \tag{4.54}$$

where N is the number of pixels of size Δx, M is the magnification, and Z is the distance from the object plane to the hologram [79].

A simple technique for extracting the z coordinate of an elongated particle image is by finding the depth plane containing the maximum voxel intensity value for each particle image cluster. In the case of two local maxima, the mean depth of the two planes provides a best estimate [80]. The in-plane particle center can then be determined using standard particle identification algorithms (discussed in section 4.1) in the identified depth plane. Satake *et al* [81] used this method of 3D particle location determination and performed an uncertainty analysis for single-camera micro holographic PTV using spherical nylon particles fixed on a glass plate. The glass plate was translated, and the measured displacements were compared to the known displacement. The in-plane and out-of-plane displacement errors were 1.8% and 11.0% of the mean velocity, respectively. These values are equivalent to displacement errors of 0.400 pixels in-plane and 2.48 pixels out-of-plane.

Due to the inaccuracy involved in depth determination techniques, many experimentalists use a stereoscopic arrangement in order to give the same accuracy for in-plane and out-of-plane position measurements [80, 82]. Stereoscopic pairing of the particle images in the two cameras is done using a nearest neighbor approach using the calculated 3D coordinates. Stereoscopic techniques require a careful calibration between the two cameras in order to achieve good results. Calibration can be performed using targets with known locations or using an experimental image itself. Lu *et al* [80] suggested using the relaxation method (discussed in section 5.3) to determine correspondences to be used for calibration. Using these particle pairs, an affine transform from the coordinates of camera 2 to the coordinates of camera 1 can be determined. Thus, all analysis can be done using the coordinate system of camera 1.

Buchmann *et al* [79] proposed using a stereoscopic holographic set-up and using tomographic reconstruction. This was done by independently reconstructing the volume imaged by both holograms, thresholding to remove background noise, then calculating the product of voxel intensities from both reconstructions. This causes any voxel with zero intensity from either hologram to become zero in the product. Thus, the uncertainty in the depth coordinate is reduced to that of the in-plane uncertainty. The error estimate for this algorithm was 0.23 pixels.

One of the more commonly used techniques of extracting 3D particle positions using a single-camera set-up is the deconvolution method, which can be applied instantaneously or iteratively to refocus a hologram to reconstruct a volume [83–88]. The instant 3D deconvolution is appropriate for the reconstruction of small tracer particles, and thus for PTV. The reconstructed object, O, is obtained from the deconvolution

$$O = \mathrm{FT}^{-1}\left[\frac{\mathrm{FT}(|\tilde{U}_r(x, y, z)|^2)}{\mathrm{FT}(|U_P|^2 + \beta)}\right], \tag{4.55}$$

where $|\tilde{U}_r(x, y, z)|^2$ is the reconstructed intensity of the object wave from convolution, $|U_P|^2$ is the reconstructed intensity of an arbitrary single point scatter, β is a small constant to avoid dividing by infinity, and FT is the 3D Fourier transform. β was set to unity in its implementation by Latychevskaia *et al* [87]. This deconvolution yields a set of sharp localizations, or ideally delta functions, that make up the particle field that was imaged. Latychevskaia and Fink [88] showed that reconstructed particle length is reduced to only two to three pixels after deconvolution.

Focus functions, which calculate the variance and Laplacian of reconstructed voxel intensities in each plane, were introduced as a method for determining the plane in which a particle is in focus [89, 90]. These techniques were still limited to selecting the reconstructed plane with the most in-focus particle image as a depth coordinate, thus Seo and Lee [91] proposed that the cross correlation of the focus functions of a tracked particle in two different frames can be used to refine the out-of-plane displacement. The cross-correlation function is defined as

$$F_1{}^{\circ}F_2 = \int F_1(z)F_2(z + \Delta z)dz, \tag{4.56}$$

where F_1 and F_2 are the focus functions in the first and second frames, respectively. Applying this algorithm to two holograms with a constant displacement of 100 μm, the standard application of the focus function was compared to the cross-correlation method. The standard deviation of measured displacement was reduced from 4.8 μm to 1.6 μm by applying the proposed cross-correlation method.

An examination of the effect of particle concentration on holographic reconstruction was performed by Kim and Lee [92]. A turbulent jet with Re = 1,200 was imaged using a PCO SensiCam CCD with 1280 × 1024 pixels in a measurement volume of 8.5 × 6.8 × 9 mm^3. The reconstruction efficiency, φ_c, was defined as the

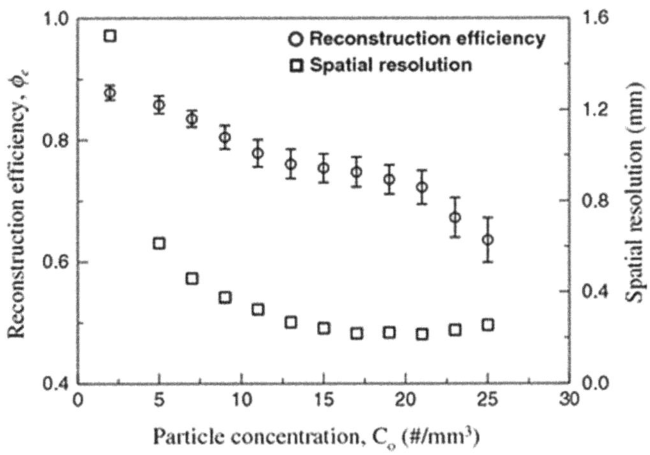

Figure 4.46. Variation of reconstruction efficiency and spatial resolution for a range of particle concentrations [92].

ratio of reconstructed particles divided by the number of seeded particles and the spatial resolution was defined as the mean particle spacing. These two values were compared for tracer concentrations varying from 2 to 25 particles/mm^3 (0.0008 to 0.0099 N_{ppp}) in figure 4.46. The reconstruction efficiency between 13 and 17 particles/mm^3 was roughly constant at 76%. Thus, the particle density corresponding to the highest spatial resolution was 17 particles/mm^3 (0.0067 N_{ppp}). This seeding density resulted in 0.0051 reconstructed N_{ppp}.

Toloui and Hong [77] proposed an iterative reconstruction technique for identifying particles in holograms called inverse iterative particle extraction (IIPE). A deconvolution is performed and particle centroids are determined as in other algorithms, then the identified particles are removed from the hologram through back-propagation. This allows particles that were hidden by the fringe patterns of identified particles to be identified. After hologram modification, the volume reconstruction, deconvolution, and centroid identification are recalculated as normal. The iterative algorithm was compared to non-iterative algorithms using synthetic data generated from 0.0008 to 0.0024 N_{ppp}. While both algorithms reconstructed more than 90% of the correct particles for the lowest seeding density tested, the iterative algorithm and non-iterative algorithms correctly reconstructed 97% and 40% of particles, respectively, for 0.0024 N_{ppp}. IIPE was further tested, successfully reconstructing at least 90% of particles up to 0.003 N_{ppp}.

Toloui *et al* [93] further improved the IIPE algorithm by using variable window local thresholding, longitudinal wall position detection, multi-pass tracking, and continuity-based displacement correction. The local thresholding algorithm initially uses a window that is six times the size of the particle diameter, reducing the window size by one particle diameter after each iteration of particle extraction. By using this technique with IIPE, the number of extractable tracers was increased by 25% for microscales and 80% for large volumes. Wall-detection was done by identifying the depth with the greatest particle concentration gradient. These

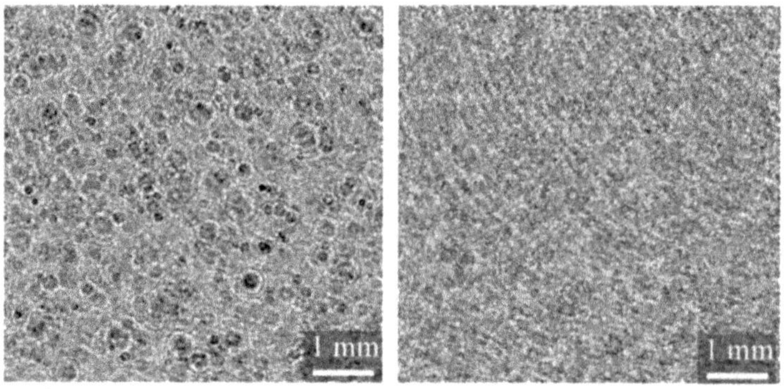

Figure 4.47. Original hologram (left) and residual hologram after using the IIPE algorithm (right). Reprinted from [92] with permission of Springer.

improvements were applied to turbulent channel flow experiments with a volume of $14.7 \times 50.0 \times 14.4$ mm^3. The maximum extracted particle density was 0.0037 N_{ppp}. A hologram that was processed using IIPE is shown in figure 4.47. Seven iterations of particle extraction left no clear particle images in the residual hologram.

Recently, Zhang *et al* [94] developed a fast and flexible MATLAB program, called UmUTracker, capable of automatically detecting and tracking particles, where images of which were captured either by light microscopy or digital in-line holographic microscopy. Validation was first performed by tracking 2D positions of polystyrene particles using both of the said imaging methods, and second 3D particle positions were assessed using synthetic and experimental holograms. The latest versions of their algorithms can be found at https://sourceforge.net/projects/umutracker/.

4.3 Optical distortions and calibration in PTV systems

In PTV systems, optical distortions exist between the image and object spaces. This causes imaging to deviate substantially from the pinhole model that is often used. Distortions can come from camera misalignment, lens imperfections and fluid–window interfaces. This section discusses how to model and compensate for these optical distortions using calibration for a single-camera 2D experiment, a planar 3D experiment, and a volumetric 3D experiment.

4.3.1 2D calibration

In order to obtain meaningful physical results from PTV, an accurate relationship between the image coordinates and the object coordinates is necessary. A standard 2D experiment uses a single camera to image particles within an illuminated sheet, which is typically one to two millimeters thick. In this case, a simplification can be made, assuming that all imaged particles have zero out-of-plane displacement from the laser light sheet. Therefore, the map from image to spatial coordinates is

$$X = F(x), \tag{4.57}$$

where X and x are the image and object coordinates, respectively. This map cannot, however, be known exactly, thus an estimate must be made. Soloff *et al* [95] suggest that a polynomial fit is sufficient in most cases and can be fitted in a least-squares sense. The authors also suggest that a third-order polynomial in each of the in-plane dimensions is sufficient for all but the most severe cases of distortion. Thus, the 2D calibration polynomial is

$$\begin{aligned}\hat{F}(x) = {} & a_0 + a_1x + a_2y + a_3x^2 + a_4xy + a_5y^2 + a_6x^3 + a_7x^2y + a_8xy^2 \\ & + a_9y^3,\end{aligned} \tag{4.58}$$

where a_i are two-component vector-valued coefficients that are determined via calibration.

Calibration is done using a target which has regularly spaced characteristic markings that cover the entire view of the camera. The characteristic markings are most often circular and must be evenly spaced on a Cartesian grid. Alternatively, Kent *et al* [96] used a single target on the end of an arm attached to a stepper motor, which was able to be moved precisely to many different locations. Before beginning the calibration procedure, coordinate systems for the image and object planes must be defined. The origin within the object coordinates is specified as the location of one specific characteristic point. All other characteristic marks are measured with respect to this position. The image plane origin is generally defined as the (0,0) pixel located in the lower-left corner of the CCD.

The first step in calibration is to acquire a calibration image. For 2D PTV, a single image is recorded with the calibration target aligned precisely at the center of the light sheet. Using the acquired image, characteristic markings are then identified using correlation-based template-matching. The correlation coefficient is determined, and a subpixel peak is identified by fitting a Gaussian or polynomial curve to the peak. Using the relative positions of these characteristic marks and their corresponding image coordinates, the coefficients of the mapping polynomial can be determined using a least-squares procedure.

Soloff *et al* [95] applied this calibration procedure using a 12×12 Cartesian grid of black circles on a white background. The circles had a diameter of 1 mm and the center-to-center spacing was 5.07 mm. The RMS calibration error was 0.54 pixels, or 0.069 mm, showing that the centroid estimation error was 7% of the particle image diameter. Prasad *et al* [97] suggested that the centroid estimation error is typically 5%–10% of the particle image diameter, demonstrating that this calibration procedure was comparable to the error contribution of centroid estimation.

4.3.2 Stereoscopic calibration

Stereo-PTV experiments generally utilize two viewing angles to determine out-of-plane displacements in a thin volume. It is important to note that some two-camera PTV set-ups make measurements in a thicker observation volume, the calibration of which should be executed using techniques discussed in section 4.3.3. In order to estimate this third component of velocity, the calibration must be performed using multiple images of a calibration target at different depths. Soloff *et al* [95] extended their proposed 2D calibration algorithm to work for planar stereoscopic measurements.

In order to calibrate a stereoscopic system, the mapping function needs to be extended to include three dimensions. The authors proposed that for a small depth, it is sufficient to only have quadratic dependence in the z direction in the calibration polynomial, therefore suggesting the polynomial

$$\begin{aligned}\hat{\boldsymbol{F}}(\boldsymbol{x}) = a_0 + a_1x + a_2y + a_3z + a_4x^2 + a_5xy \\ + a_6y^2 + a_7xz + a_8yz + a_9z^2 + a_{10}x^3 + a_{11}x^2y \\ + a_{12}xy^2 + a_{13}y^3 + a_{14}x^2z \\ + a_{15}xyz + a_{16}y^2z + a_{17}xz^2 + a_{18}yz^2.\end{aligned} \tag{4.59}$$

The quadratic dependence of z requires a minimum of three calibration images at different depths, although more calibration images can be used to reduce the mean squared error of the estimate. After the image coordinates of all of the characteristic marks in the calibration images have been determined, the calibration polynomial coefficients can be determined for each of the two cameras separately.

Soloff *et al* [95] tested their calibration technique using a target made of a rectangular block of aluminum with 0.5 mm holes drilled in a 9×9 Cartesian grid. The center-to-center spacing was 9 mm between holes. The calibration images were taken of the backlit plate positioned at the center of the light sheet and with a 0.5 mm offset in both directions. The RMS errors for the mapping function were 1.1 and 1.2 pixels, or 0.045 and 0.051 mm, for cameras one and two, respectively. The out-of-plane displacement estimates based on a PIV interrogation technique were less than 0.02 mm for a measurement depth of 1 mm.

Wieneke [98] introduced a calibration technique for stereoscopic imaging that use ray-tracing corrected for distortions (see section 4.2.1.2) and does not require the calibration plate to be aligned with the laser sheet. This allows for calibration to be performed even if the observation volume cannot be easily accessed. In addition, this approach uses tracer particle images rather than images of a calibration target, and is therefore called the self-calibration technique. The first step of the self-calibration technique is to determine a 3D mapping between image and object coordinates, either using the technique suggested by Soloff *et al* [95] or the ray-tracing procedure, for example, the distortion-corrected ray-tracing proposed by Wieneke (section 4.2.1.2). The second step is to compute the disparity vector map of the two dewarped images collected by each of the two cameras. This is done using standard cross-correlation. In order to average out the effect of depth, between 5 and 50 image pairs should be used in an ensemble-averaging algorithm. A multi-pass algorithm can use this initial disparity vector map to distort the cross-correlation interrogation windows, which can then be used for another cross-correlation to further refine the disparity estimate. Once this disparity map has been calculated, corresponding object coordinates are determined for each vector using triangulation. The disparity vectors mean that the two re-projected lines from each camera do not intersect at a single point in the object plane. Therefore, triangulation must be used to determine the point in space that minimizes the disparity between what was measured in camera one and camera two. This triangulation using disparity vectors is shown in figure 4.48.

Once the triangulation has been performed for each disparity vector, a plane is then fitted to the world points in 3D space and the mapping functions of both cameras are corrected by a corresponding transformation such that the new measurement plane becomes the $z = 0$ plane. The translation and rotation of the new measurement plane are finally applied to the original mapping functions for both cameras. The whole process can be repeated again to arrive at better fits, and Wieneke suggested that using single-pass cross-correlation and repeating the entire correction procedure two or three times converges to gives good results.

It was experimentally shown that this self-calibration procedure was able to place the $z = 0$ plane within 0.1 pixels of the center of the light sheet. It is advisable to use this procedure to check for the proper alignment of the calibration plate. It was also proposed that calibration could be performed for a closed measurement volume using self-calibration. The cameras must be placed on a traverse such that they can move away from the measurement volume. They are focused on the measurement plane and traversed backward until a calibration target mounted on the front side of the measurement volume is in focus. The initial calibration is done in air and the ray-tracing technique is then used to correct for the distorting media. For this step, the distance between the calibration target and the laser sheet must be known. Self-calibration then corrects further disparities when the cameras have been moved back to observe the measurement plane. Using stereo-PIV, out-of-plane displacements were measured to within 2% for externally calibrated systems using self-calibration. This approach has not been extended into PTV analyses; therefore, it is unknown how this uncertainty carries into PTV results.

The *in situ* calibration technique was used by Choi and Guezennec [99] to calibrate stereo-PTV systems that use a wide-angle separation between cameras to

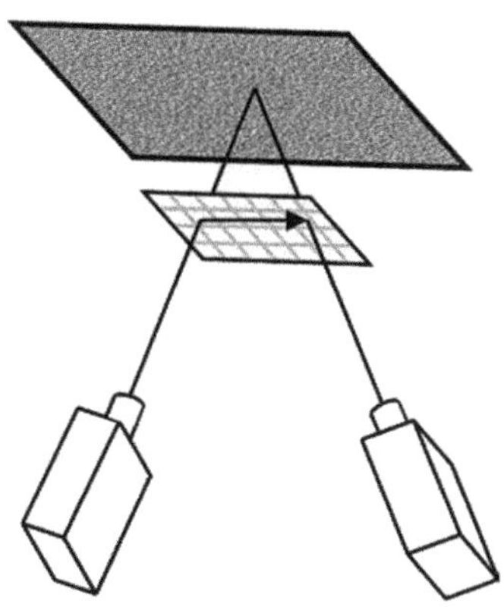

Figure 4.48. The object points of the light sheet determined using a disparity vector [98].

image a deep volume. It was proposed that four polynomials should be fitted to fully describe the relationship between image and object coordinates for two cameras. For camera two, a particle image located at $(x_2.y_2)$ corresponds to world coordinates

$$x = f_2(x_2, y_2. y), \tag{4.60}$$

$$z = g_2(x_2, y_2. y), \tag{4.61}$$

where y is a free variable and f_2 and g_2 are polynomial functions. The corresponding relations for camera one are

$$y = f_1(x_1, y_1. x), \tag{4.62}$$

$$z = g_1(x_1, y_1. x), \tag{4.63}$$

where x is a free variable and f_1 and g_1 are polynomial functions. The coefficients of the polynomial functions can be obtained using calibration points and least-squares curve-fitting. The authors suggested that a second-order fit in all three variables was sufficient to achieve good model accuracy. This calibration technique was similarly shown to outperform ray-tracing when geometrical–optical parameters were not known exactly.

4.3.3 3D calibration

In the context of 3D-PTV, an accurate calibration procedure plays an important role not only for minimizing errors, but also for reconstructing true particles without the presence of ghosts. 3D-PTV techniques require a mapping function between a camera's image space and the object space. While photogrammetric techniques define the object space in terms of physical coordinates, tomographic, plenoptic, and synthetic aperture systems define the object space in discrete voxels.

The volumetric calibration procedure is largely the same as that of a thin sheet, except that the calibration target needs to be imaged throughout the entire volume for the most accurate results. While the standard calibration target is either a metal sheet with machined holes or dark dots on a white background, Schosser *et al* [100] showed that a non-intrusive calibration can be performed by creating calibration points using a laser. The proposed technique uses a continuous wave laser with a focused beam, perpendicularly aligned to the optical access window to illuminate the measurement volume at known x and y locations. Whenever the light impacts a surface, it induces light reflections, which can be imaged and used as calibration points. The reflections from the front and rear boundaries of the measurement volume can be used for the calibration, as shown in figure 4.49. This non-intrusive method of obtaining calibration images within a Tesla turbine rotor demonstrates the usefulness of this technique for performing 3D velocimetry when measurement volume access is not feasible, such as in turbomachinery, cavity flows, and combustion.

The calibration points are imaged on each camera and used to calculate the parameters of the mapping function. There are generally two different types of

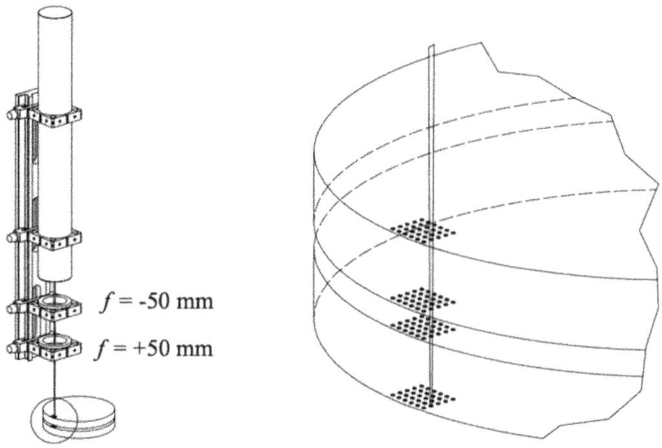

Figure 4.49. Calibration points obtained using laser reflection from measurement volume boundaries. Reprinted from [100] with permission of Springer.

mapping function that can be selected: one that models the refractive interfaces for a given experimental set-up, and a general one that aims to account for all potential sources of distortion. A common way to model an experimental set-up is to use the pinhole model and correct it for distortions. The equations describing the transformation from image to epipolar line are given in section 4.2.1.2. The unknowns can be found by performing a least-squares fit to the calibration images [101].

Maas [102] suggested using a correction to model for refractive index changes in the line-of sight for particle tracking experiments. More recently Mulsow [103] incorporated a camera system model with refraction using ray tracing. This approach was applied to planar interfaces and cylindrical vessels by Belden [61]. Generally speaking, ray-tracing states that a point in space in the direction of ray $\boldsymbol{r}$ is given by

$$\boldsymbol{X}(t) = \boldsymbol{X}_0 + t\boldsymbol{r}, \tag{4.64}$$

where $\boldsymbol{X}_0$ is a known point on the line and t is a parameter that determines the location along the line. The unit vector that defines the direction of $\boldsymbol{r}$, the refracted ray, is given by

$$\hat{\boldsymbol{r}}_t = n_{it}\hat{\boldsymbol{r}}_i + \left[-n_{it}\hat{\boldsymbol{N}} \cdot \hat{\boldsymbol{r}}_i - \sqrt{1 - n_{it}^2[1 - (-\hat{\boldsymbol{N}} \cdot \hat{\boldsymbol{r}}_i)]}\right]\hat{\boldsymbol{N}}, \tag{4.65}$$

where n_{it} is the ratio of refractive indices of the incident and transmitted media, $\hat{\boldsymbol{N}}$ is the unit normal of the refractive interface, and $\hat{\boldsymbol{r}}_i$ is the initial ray direction. Using these refraction equations, the alternating forward ray-tracing (AFRT) procedure, shown in figure 4.50, begins by assuming a straight ray from the camera to the object and iteratively determining the correspondence between object and image coordinates. Due to the direct modeling of the refraction, the standard deviation of position estimates in a simulation was found to be a length equivalent to 0.1 pixels.

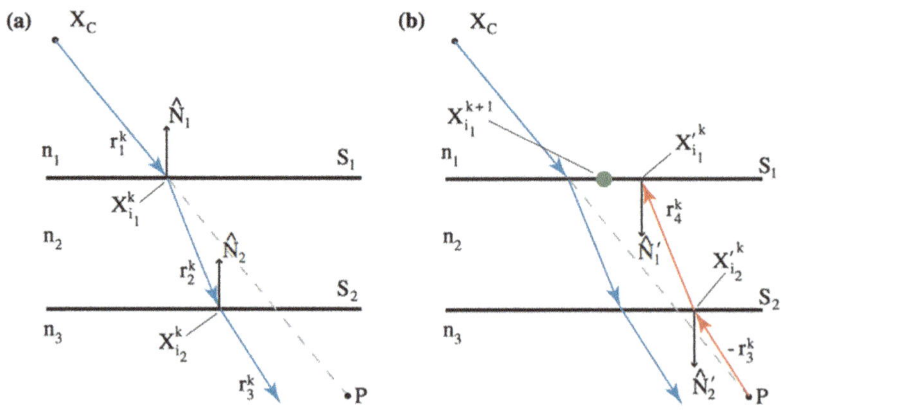

Figure 4.50. Schematic of the AFRT procedure used by Belden. P is the point of interest and X_c is the image coordinate. Reprinted from [100] with permission of Springer.

A model-free calibration method, first suggested by Kent and Trigui [104], suggests using an optical transform with coefficients determined by *in situ* calibration instead of ray-tracing techniques, which require accurate knowledge of the geometry of the experimental set-up. In their set-up, Kent and Trigui show that their calibration technique outperforms ray tracing. To show this, an experimental set-up with two cameras oriented at 90 degrees to one another was simulated. Camera 1 was given various azimuthal offsets as well as a 5 degree offset in elevation. The comparison of Kent and Trigui's *in situ* calibration technique and traditional ray tracing for this camera misalignment and zero 2D error is shown in figure 4.51. The errors for *in situ* calibration are two orders of magnitude less than those of ray tracing, suggesting that camera misalignment is not a major source of error for the calibration technique. As a result, errors from 2D localization cause the most substantial uncertainty for this method.

This model-free calibration technique was also used by Tien *et al* [105] for calibration in the presence of nonhomogeneous media and optical imperfections. In the calibration process, a transformation matrix between image coordinates and object coordinates is found for each z coordinate used. Then for each pixel in the camera, the spatial x and y coordinates for each z coordinate are calculated using this transformation. The technique uses radial basis functions to approximate the mapping function. The calibration error estimates from these experiments in the context of defocusing micro-PIV are below 4 microns (1.5% of calibration range) throughout the measurement volume. The in-plane RMS errors are less than 0.16 microns (0.05%) and the out-of plane errors are around 0.81 microns (0.25%).

Yoon *et al* [27] found that in the context of defocusing micro-PIV, experimental calibration is necessary for determining the depth of a particle. A relationship between the depth calibration coefficient, $\partial z/\partial D$, and the fluid refractive index, RI, was presented by the authors, which allowed for a single calibration to be used for a wide range of fluids given the same experimental set-up. The depth calibration coefficient gives an estimate of the particle distance from the focal plane using the

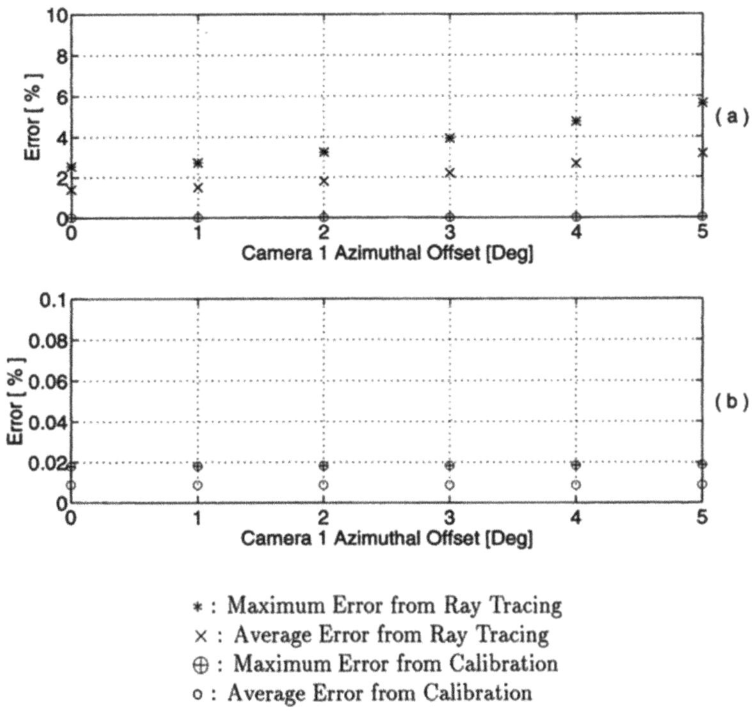

Figure 4.51. 3D positioning errors from camera azimuthal misalignment. Reprinted from [61] with permission of Springer.

diameter of the circle connecting the three defocused particle images. The proposed estimate of the calibration coefficient is

$$(\partial z/\partial D)_{\text{fluid}} \approx \frac{n_{\text{fluid}}}{n_{\text{ref}}}(\partial z/\partial D)_{\text{ref}}, \tag{4.66}$$

where n_{fluid} and n_{ref} are the refractive indices of an experimental fluid and a reference fluid, respectively. This is an approximate relationship that is valid for small angles of refraction. Applying this technique experimentally, the calibration coefficient only had discrepancies on the order of 1% from the directly measured coefficient.

An additional volume self-calibration (VSC) step can be included using only the collected triangulation data. Wieneke [106] suggested that the residual triangulation error can be used to correct the mapping functions for each camera. 2D particle image positions, x_i and y_i, are mapped to their best fit 3D spatial coordinates. Then these 3D coordinates are mapped back to image coordinates, x_i' and y_i', and a disparity, d_i, exists for real data. Typical self-calibration techniques adjust their mapping function, M_i, by this disparity distribution,

$$d_i = (x_i', y_i') - (x_i, y_i) \tag{4.67}$$

$$M_i'(X, Y, Z) = M_i(X, Y, Z) - d_i(X, Y, Z). \tag{4.68}$$

This technique works well for low particle seeding densities when particle correspondences can be matched with a high confidence level and few ghost particles. Wieneke proposed that one can apply a threshold such that only the brightest 10% of particle images are considered. After thresholding, the volume is subdivided such that multiple particle images exist in each. The disparity is then plotted for all particles in a sub-volume for each camera and the peak is selected. The disparities are then smoothed throughout the entire volume and used to correct the mapping function of each camera. The procedure can be repeated for further improvement and validation. In the context of tomographic PIV with 0.1 N_{ppp}, the maximum calibration errors were shown to be reduced from 1–2 to 0.1–0.2 pixels in some areas. This error is due to the disparity from the VSC triangulation step, and thus carries into PTV analyses as well. By recognizing that true particles always have disparities for all the cameras within the true disparity peak and that ghost particles are randomly dispersed about the disparity peak for each camera, Wieneke was able to improve the VSC algorithm (VSC-GPS) to reduce the number of ghost particles by many orders of magnitude [107].

Discetti and Astarita [108] investigated the performance of this self-calibration algorithm and determined that including four or more cameras in the procedure can significantly affect the quality of the disparity maps due to the contribution of spurious matches. It was proposed that VSC should first be performed on three cameras, then again for each additional camera using the first three as a reference. This method is recommended for cases with high particle seeding densities and a large number of cameras. It was shown with simulated data at 0.05 N_{ppp} that the modified self-calibration algorithm can improve the signal-to-noise ratio, particularly when a search radius greater than five pixels is required for triangulation and more than four cameras are used.

Another calibration approach used by tomo-PTV and IPR is the non-uniform optical transfer function (OTF) [109]. The technique uses a spatially variable correction for distorted particle image shapes in order to improve reconstruction results. The OTF algorithm relies upon the VSC results for 3D coordinates of particles and the corresponding image coordinates on each camera. These images are each characterized in terms of the 2D Gaussian peak-fitter,

$$W(x, y) = p * \exp\left(-\frac{1}{2}\begin{pmatrix} x \\ y \end{pmatrix}^T \begin{pmatrix} a & b \\ b & c \end{pmatrix}\begin{pmatrix} x \\ y \end{pmatrix}\right), \tag{4.69}$$

where p, a, b, and c describe the shape and peak height of a particle image. These image characteristics are collected and separated into sub-volumes. Within these sub-volumes the image characteristics are averaged for each camera, allowing for an optical transformation that varies in space. The volume can be discretized into any sized sub-volumes in order to best account for spatial variations (figure 4.52).

Both a simulation and an experimental test of the OTF calibration were performed. For both perfect and astigmatic data, SMART reconstruction was performed using OTF, a bilinear interpolation, and a B-spline for weighting functions. These were all compared to a MART algorithm with no optical

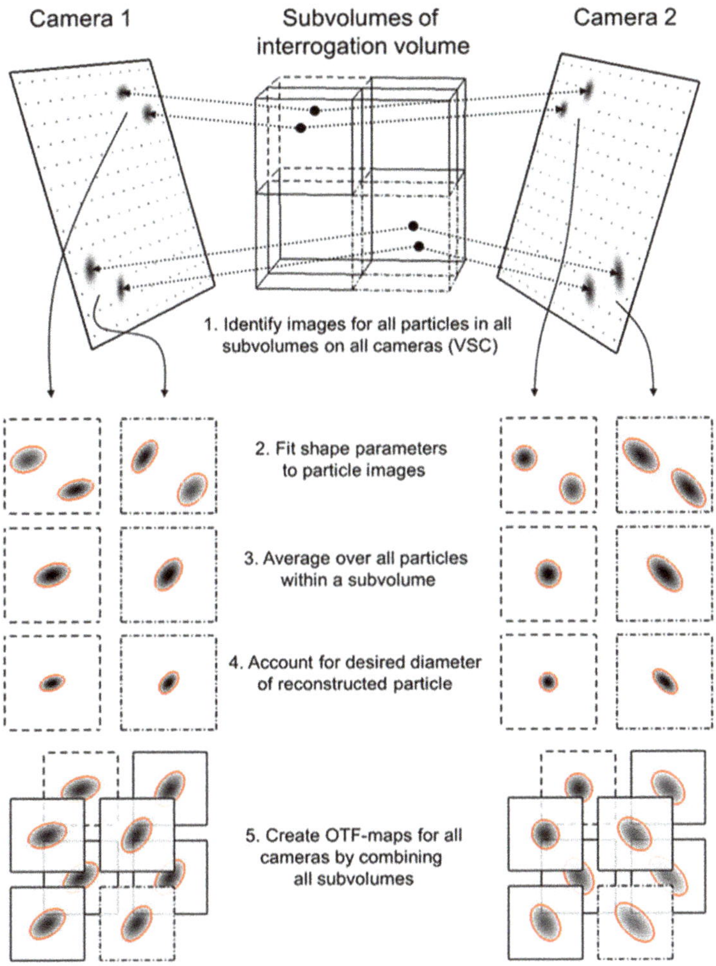

Figure 4.52. Steps required for an OTF calibration [109].

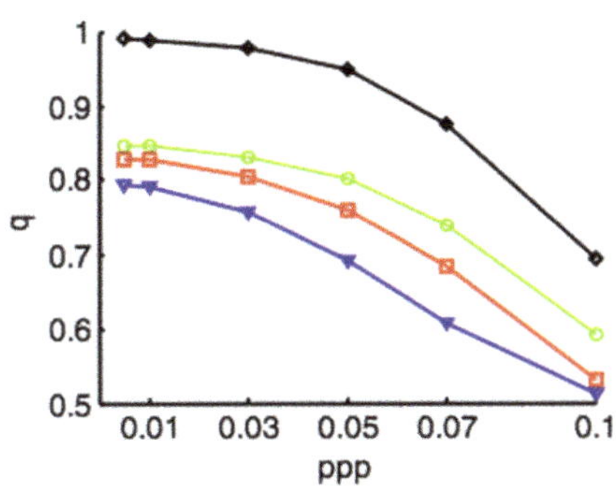

Figure 4.53. Comparison of reconstruction quality for OTF (black), B-spline (green), bilinear interpolation (red), and MART (blue) using synthetic astigmatic data [109].

calibration, and in all cases tested up to 0.1 N_{ppp}, where the OTF outperformed the others in terms of reconstruction quality, q, as shown in figure 4.53.

Also important to note are the self-calibration methods developed for Scheimflug cameras incorporating different camera models that can be used for stereo or 3D-PTV [110–113]. While Cornic *et al* [112] do point out that conventional methods provide proper calibration for small lens tilt angles ($\leqslant 6°$), they also point out that for larger angles, the calibration methods must incorporate Scheimflug cameras in order to adequately calibrate the error due to tangential distortion. A review of these approaches is provided by Sun *et al* [114].

References

[1] Adrian R J 1991 Particle-imaging techniques for experimental fluid mechanics *Annu. Rev. Fluid Mech.* **23** 261–304

[2] Prasad A K 2000 Stereoscopic particle image velocimetry *Exp. Fluids* **29** 103–16

[3] Watanabe R, Gono T, Yamagata T and Fujisawa N 2015 Three-dimensional flow structure in a highly buoyant jet by scanning stereo PIV combined with POD analysis *Int. J. Heat Fluid Flow* **52** 98–110

[4] Doh D H, Hwang T G and Saga T 2004 3D-PTV measurements of the wake of a sphere *Meas. Sci. Technol.* **15** 1059–66

[5] Graff E C and Gharib M 2008 Performance prediction of point-based three-dimensional volumetric measurement systems *Meas. Sci. Technol.* **19** 075403

[6] Kreizer M and Liberzon A 2011 Three-dimensional particle tracking method using FPGA-based real-time image processing and four-view image splitter *Exp. Fluids* **50** 613–20

[7] Murai S, Nakamura H and Suzuki Y 1980 Analytical orientation for non-metric camera in the application to terrestrial photogrammetry *Int. Arch. Photogrammetry* **23** 516–25

[8] Nishina K, Kasagi N and Hirata M 1989 Three-dimensional particle tracking velocimetry based on automated digital image processing *Trans. ASME* **111** 384–91

[9] Maas H G, Gruen A and Papantoniou D 1993 Particle tracking velocimetry in three-dimensional flows Part 1. Photogrammetric determination of particle coordinates *Exp. Fluids* **15** 133–46

[10] Brown D 1971 Close-range camera calibration *Photogrammetric Eng.* **37** 855–66

[11] El-Hakrm S F 1986 Real-time image metrology with CCD cameras. *Photogramm. Eng. Remote Sens.* **52** 1757–66

[12] Albertz J and Kreiling W 1989 *PhotogrammetrischeTaschenbuch* (Karlsruhe: Wichmann)

[13] Lei Y-C, Tien W-H, Duncan J, Paul M, Ponchaut N, Mouton C, Dabiri D, Rösgen T and Hove J 2012 A vision-based hybrid particle tracking velocimetry (PTV) technique using a modified cascade-correlation peak-finding method *Exp. Fluids* **53** 1251–68

[14] Tien W H, Kartes P, Yamasaki T and Dabiri D 2008 A color-coded backlighted defocusing particle image velocimetry system *Exp. Fluids* **44** 1015–26

[15] Tien W H, Dabiri D and Hove J R 2014 Color-coded three-dimensional micro particle tracking velocimetry and application to micro backward-facing step flows *Exp. Fluids* **55** 1684

[16] Tien W-H and Dabiri D 2012 Color-coded three-dimensional micro particle tracking velocimetry and application to micro accelerating micro-channel flow *Int. Symp. on Applications of Laser Techniques to Fluid Mechanics (Lisbon, Portugal, 9–25 July)*

[17] Fuchs T, Hain R and Kähler C J 2016 Double-frame 3D-PTV using a tomographic predictor *Exp. Fluids* **57** 174
[18] Wieneke B 2013 Iterative reconstruction of volumetric particle distribution *Meas. Sci. Technol.* **24** o24008
[19] Bown M R, MacInnes J M, Allen R W K and Zimmerman W B J 2006 Three-dimensional, three-component velocity measurements using stereoscopic micro-PIV and PTV *Meas. Sci. Technol.* **17** 2175–85
[20] Yu C H, Yoon J H and Kim H B 2009 Development and validation of stereoscopic micro-PTV using match probability *J. Mech. Sci. Technol.* **23** 845–55
[21] Peterson K, Regaard B, Heinemann S and Sick V 2012 Single-camera, three-dimensional particle tracking velocimetry *Opt. Express* **20** 9031–37
[22] Bao X and Li M 2011 Defocus and binocular vision based stereo pairing method for 3D particle tracking velocimetry *Opt. Lasers Eng.* **49** 623–31
[23] Panday S P, Ohmi K and Nose K 2011 An ant colony optimization based stereoscopic particle pairing algorithm for three-dimensional particle tracking velocimetry *Flow Meas. Instrum.* **22** 86–95
[24] Hoyer K, Holzner M, Lüthi B, Guala M, Liberzon A and Kinzelbach W 2005 3D scanning particle tracking velocimetry *Exp. Fluids* **39** 923–34
[25] Willert C E and Gharib M 1992 3-dimensional particle imaging with a single camera *Exp. Fluids* **12** 353–8
[26] Pereira F, Gharib M, Dabiri D and Modarress D 2000 Defocusing digital particle image velocimetry: a 3-component 3-dimensional DPIV measurement technique. Application to bubbly flows *Exp. Fluids* **29** S78–84
[27] Yoon S Y and Kim K C 2006 3D particle position and 3D velocity field measurement in a microvolume via the defocusing concept *Meas. Sci. Tech.* **17** 2897–905
[28] Pereira F, Lu J, Castano-Graff E and Gharib M 2007 Microscale 3D flow mapping with μDDPIV *Exp. Fluids* **42** 589–99
[29] Tien W-H 2016 Development of multi-spectral three-dimensional micro particle tracking velocimetry *Meas. Sci. Tech.* **27** 084010
[30] Kajitani L and Dabiri D 2005 A full three-dimensional characterization of defocusing digital particle image velocimetry *Meas. Sci. Technol.* **16** 790–804
[31] Grothe R L and Dabiri D 2008 An improved three-dimensional characterization of defocusing digital particle image velocimetry (DDPIV) based on a new imaging volume definition *Meas. Sci. Technol.* **19** 065402
[32] Pereira F and Gharib M 2002 Defocusing digital particle image velocimetry and the three-dimensional characterization of two-phase flows *Meas. Sci. Technol.* **13** 683–94
[33] Kao H P and Verkman A S 1994 Tracking of single fluorescent particles in three dimensions cylindrical optics to encode particle position *Biophys. J.* **67** 1291–300
[34] Towers C E, Towers D P, Campbell H I, Zhang S and Greenaway A H 2006 Three-dimensional particle imaging by wavefront sensing *Opt. Lett.* **31** 1220–2
[35] Angarita-Jaimes N, McGhee E, Chennaoui M, Campbell H I, Zhang S, Towers C E, Greenaway A H and Towers D P 2006 Wavefront sensing for single view three-component three-dimensional flow velocimetry *Exp. Fluids* **41** 881–91
[36] Chen S, Angarita-Jaimes N, Angarita-Jaimes D, Pelc B, Greenaway A H, Towers C E, Lin D and Towers D P 2009 Wavefront sensing for three-component three-dimensional flow velocimetry in microfluidics *Exp. Fluids* **47** 849–63

[37] Cierpka C, Segura R, Hain R and Kähler C J 2010 A simple single camera 3D3C velocity measurement technique without errors due to depth of correlation and spatial averaging for microfluidics *Meas. Sci. Technol.* **21** 045401

[38] Cierpka C, Rossi M, Segura R and Kähler C J 2011 On the calibration of astigmatism particle tracking velocimetry for microflows *Meas. Sci. Tech.* **22** 015401

[39] Rossi M and Kähler C J 2014 Optimization of astigmatic particle tracing velocimeters *Exp. Fluids* **55** 1809

[40] Fuchs T, Hain R and Kähler C J 2014 Three-dimensional location of micrometer-sized particles in macroscopic domains using astigmatic aberrations *Opt. Lett.* **39** 5

[41] Fuchs T, Heinold M, Hain R and Kähler C J 2014 Stereoscopic astigmatism particle tracking velocimetry for macroscopic 3D3C flow measurements in air *17th Int. Symp. on Applications of Laser Techniques to Fluid Mechanics (Lisbon, Portugal)*

[42] Fuchs T, Hain R and Kähler C J 2014 Macroscopic three-dimensional particle location using stereoscopic imaging and astigmatic aberrations *Opt. Lett.* **39** 24

[43] Cagnet M, Francon M and Thrierr J C 1962 *Atlas of Optical Phenomena* (Berlin: Springer)

[44] Gibson S F and Lanni F 1991 Experimental test of an analytical model of aberration in an oil-immersion objective lens used in three-dimensional light microscopy *J. Opt. Soc. Am.* A **8** 1601–13

[45] Park J S Study of microfluidc measurement techniques using novel optical imaging diagnostics *PhD Dissertation, Texas A&M University, College Station, TX*

[46] Speidel M, Jonás A and Florin E-L 2003 Three-dimensional tracking of fluorescent nanoparticles with subnanometer precision by use of off-focus imaging *Opt. Lett.* **28** 69–71

[47] Wu M, Roberts J W and Buckley M 2005 Three-dimensional fluorescent particle tracking at micron-scale using a single camera *Exp. Fluids* **38** 461–5

[48] Park J S and Kihm K D 2006 Three-dimensional micro-PTV using deconvolution microscopy *Exp. Fluids* **40** 491–9

[49] Luo R, Yang X Y, Peng X F and Sun Y F 2006 Three-dimensional tracking of fluorescent particles applied to micro-fluidic measurements *J. Micromech. Microeng.* **16** 1689–99

[50] Haeberlé O 2003 Focusing of light through a stratified medium: a practical approach for computing microscope point spread functions: Part I. Conventional microscopy *Opt. Commun.* **216** 55–63

[51] Luo R, Sun Y F, Peng X F and Yang X Y 2006 Tracking sub-micron fluorescent particles in three dimensions with a microscope objective under non-design optical conditions *Meas. Sci. Tech.* **17** 1358–66

[52] Luo R and Sun Y F 2011 Pattern matching for three-dimensional tracking of sub-micron fluorescent particles *Meas. Sci. Tech.* **22** 045402

[53] Peterson S D, Chuang H-S and Wereley S T 2008 Three-dimensional particle tracking using micro-particle image velocimetry hardware *Meas. Sci. Technol.* **19** 115406

[54] Elsinga G E, Scarano F L, Wieneke B and van Oudheusden B W 2006 Tomographic particle image velocimetry *Exp. Fluids* **41** 933–47

[55] Doh D H, Cho G R and Kim Y H 2012 Development of a tomographic PTV *J. Mech. Sci. Technol.* **26** 3811–9

[56] Atkinson C and Soria J 2009 An efficient simultaneous reconstruction technique for tomographic particle image velocimetry *Exp. Fluids* **47** 553–68

[57] Kitzhofer J and Brücker C 2010 Tomographic particle tracking velocimetry using telecentric imaging *Exp. Fluids* **49** 1307–24

[58] Belden J, Truscott T T, Axiak M C and Techet A H 2010 Three-dimensional synthetic aperture particle image velocimetry *Meas. Sci. Technol.* **21** 125403
[59] Bajpayee A and Techet A H 2013 3D particle tracking velocimetry (PTV) using high speed light field imaging *PIV13; 10th Int. Symp. on Particle Image Velocimetry (Delft, The Netherlands, July 2013)*
[60] Vaish V *et al* 2005 Synthetic aperture focusing using a shear-warp factorization of the viewing transform *IEEE Computer Society Conf. on Computer vision and Pattern Recognition Workshops* pp 129–36
[61] Belden J 2013 Calibration of multi-camera systems with refractive interfaces *Exp. Fluids* **54** 1–18
[62] Bajpayee A and Techet A H 2017 Fast volume reconstruction for 3D PTV *Exp. Fluids* **58** 95
[63] Ng R 2006 Digital light field photography *PhD Thesis* Stanford University, CA
[64] Adelson E H and Wang J Y A Single lens stereo with a plenoptic camera *IEEE Trans. Pattern Anal. Mach. Intell.* **14** 99–106
[65] Ng R, Levoy M, Duval G, Horowitz M and Hanrahan P 2005 Light field photography with a hand-held plenoptic camera *Technical Report* CSTR 2005-02 Stanford University, CA http://graphics.stanford.edu/papers/lfcamera/
[66] Lumsdaine A and Georgiev T 2009 The focused plenoptic camera *Proc. IEEE Int. Conf. Comput. Photogr., Apr. 2009* pp 1–8
[67] Fahringer T, Lynch K and Thurow B 2015 Volumetric particle image velocimetry with a single plenoptic camera *Meas. Sci. Tech.* **26** 115201
[68] Fahringer T and Thurow B 2013 The effect of grid resolution on the accuracy of tomographic reconstruction using a plenoptic camera *Proc. 51st AIAA Aerosp. Sci. Meeting Including New Horizons Forum Aerosp. Expo.* http://doi.org/10.2514/6.2013–39
[69] Fahringer T and Thurow B 2016 Filtered refocusing: a volumetric reconstruction algorithm for plenoptic-PIV *Meas. Sci. Tech.* **27** 094005
[70] Anglin P, Reeves S J and Thurow B S 2017 Characterization of plenoptic imaging systems and efficient volumetric estimation from plenoptic data *IEEE J. Selected Topics Signal Process.* **11** 1020–33
[71] Chen H and Sick V 2017 Three-dimensional three-component air flow visualization in a steady-state engine flow bench using a plenoptic camera *SAE Int. J. Engines* **10** 625–35
[72] Pu Y and Meng H 2006 An advanced off-axis holographic particle image velocimetry (HPIV) system *Exp. Fluids* **29** 184–97
[73] Yu X, Hong J, Liu C and Kim M K 2014 Review of digital holographic microscopy for three-dimensional profiling and tracking *Opt. Eng.* **53** 112306
[74] Memmolo P, Miccio L, Paturzo M, Di Caprio G, Coppola G, Netti P A and Ferraro P 2015 Recent advances in holographic 3D particle tracking *Adv. Optics Photonics* **7** 713
[75] Adrian R J and Westerweel J 2011 *Particle Image Velocimetry* (New York: Cambridge University Press)
[76] Toloui M, Mallery K and Hong J 2017 Improvements on digital inline holographic PTV for 3D wall-bounded turbulent flow measurements *Meas. Sci. Technol.* **28** 044009
[77] Toloui M and Hong J 2015 High fidelity digital inline holographic method for 3D flow measurements *Opt. Express* **23** 1364
[78] Choi Y S, Seo K W, Sohn M H and Lee S J 2012 Advances in digital holographic micro-PTV for analyzing microscale flows *Opt. Lasers Eng.* **50** 39–45
[79] Buchmann N A, Atkinson C and Soria J 2013 Ultra-high-speed tomographic digital holographic velocimetry in supersonic particle-laden jet flows *Meas. Sci. Technol.* **24** 024005

[80] Lu J, Fuga J P, Nordsiek H, Saw E W, Shaw R A and Yang W 2008 Lagrangian particle tracking in three dimensions via single-camera in-line digital holography *New J. Phys.* **10** 125013
[81] Satake S, Anraku T, Kanamori H, Kunugi T, Sato K and Ito T 2008 Measurements of three-dimensional flow in microchannel with complex shape by micro-digital-holographic particle-tracking velocimetry *J. Heat Transfer* **130** 042412-1
[82] Schablinski J, Kroll M and Block D 2013 Particle tracking velocimetry of dusty plasmas using stereoscopic in-line holography *IEEE Trans. Plasma Sci.* **41** 779
[83] Shaw P J 1995 Comparison of wide-field/deconvolution and confocal microscopy for 3D imaging *Handbook of Biological Confocal Microscopy* ed J B Pawley pp 373–87 (New York: Plenum)
[84] McNally J G *et al* 1999 Three-dimensional imaging by deconvolution microscopy *Methods* **19** 373–85
[85] Sarder P and Nehorai A 2006 Deconvolution methods for 3-D fluorescence microscopy images *IEEE Signal Process Mag.* **23** 32–45
[86] Wallace W, Schaefer L H and Swedlow J R 2001 A working person's guide to deconvolution in light microscopy *Biotechniques* **31** 1076–97
[87] Latychevskaia T, Gehri F and Fink H 2010 Depth-resolved holographic reconstructions by three-dimensional deconvolution *Opt. Express* **18** 22527–44
[88] Latychevskaia T and Fink H W 2014 Holographic time-resolved particle tracking by means of three-dimensional volumetric deconvolution *Opt. Express* **22** 1364
[89] Choi Y S and Lee S J 2009 Three-dimensional volumetric measurement of red blood cell motion using digital holographic microscopy *Appl. Opt.* **48** 2983–90
[90] Lee S J, Seo K W, Choi Y S and Sohn M H 2011 Three-dimensional motion measurements of free-swimming microorganisms using digital holographic microscopy *Meas. Sci. Tech.* **22** 064004
[91] Seo K W and Lee S J 2014 High-accuracy measurement of depth-displacement using a focus function and its cross-correlation in holographic PTV *Opt. Express* **22** 1364
[92] Kim S and Lee S J 2008 Effect of particle number density in in-line digital holographic particle velocimetry *Exp. Fluids* **44** 623–31
[93] Toloui M, Mallery K and Hong J 2017 Improvements on digital inline holographic PTV for 3D wall-bounded turbulent flow measurements *Meas. Sci. Technol.* **28** 044009
[94] Stangner T, Wiklund K, Rodriguez A and Andersson M 2017 UmUTracker: a versatile MATLAB program for automated particle tracking of 2D light microscopy or 3D digital holography data *Comput. Phys. Commun.* **219** 390–9
[95] Soloff S M, Adrian R J and Liu Z C 1997 Distortion compensation for generalized stereoscopic particle image velocimetry *Meas. Sci. Technol.* **8** 1441–54
[96] Kent J C, Trigui N, Choi W-C, Guezennec Y G and Brodkey R S 1993 Photogrammetric calibration for improved three-dimensional particle tracking velocimetry *Optical Diagnostics in Fluid and Thermal Flow* ed S Cha and J Trolinger (Bellingham, WA: SPIE) pp 400–12
[97] Prasad A K, Adrian R J, Landreth C C and Offutt P W 1992 Effect of resolution on the speed and accuracy of particle image velocimetry interrogation *Exp. Fluids* **13** 105–16
[98] Wieneke B 2005 Stereo-PIV using self-calibration on particle images *Exp. Fluids* **39** 267–80
[99] Choi W and Guezennec Y 1994 *In situ* calibration for wide-angle, three-dimensional stereoscopic image analysis *Appl. Opt.* **36** 7364–73

[100] Schosser C, Fuchs T, Hain R and Kahler C J 2016 Non-intrusive calibration for three-dimensional particle imaging *Exp. Fluids* **57** 69
[101] Maas H G, Gruen A and Papantoniou D 1993 Particle tracking velocimetry in three-dimensional flows Part 1. Photogrammetric determination of particle coordinates *Exp. Fluids* **15** 133–46
[102] Maas H G 1995 New developments in multimedia photogrammetry ed A Gruñ and H Kahmen *Optical 3-D measurement techniques III* (Karlsruhe: Wichmann)
[103] Mulsow C 2010 A flexible multi-media bundle approach *Int. Arch. Photogrammetry* **38** 472–7
[104] Kent J C and Trigui N 1993 Photogrammetric calibration for improved three dimensional particle tracking velocimetry (3DPTV) *Proc. SPIE* **2005** 400–12
[105] Tien W H, Dabiri D and Hove J R 2014 Color-coded three-dimensional micro particle tracking velocimetry and application to micro backward-facing step flows *Exp. Fluids* **55** 1684
[106] Wieneke B 2008 Volume self-calibration for 3D particle image velocimetry *Exp. Fluids* **45** 549–56
[107] Wieneke B 2018 Improvements for volume self-calibration *Meas. Sci. Technol.* **29** 084002
[108] Discetti S and Astarita T 2014 The detrimental effect of increasing the number of cameras on self-calibration for tomographic PIV *Meas. Sci. Technol.* **25** 084001
[109] Schanz D, Gesemann S, Schroder A, Wieneke B and Novara M 2013 Non-uniform optical transfer functions in particle imaging: calibration and application to tomographic reconstruction *Meas. Sci. Technol.* **24** 024009
[110] Fournel T, Lavest J M, Coudert S and Collange F 2004 Self-calibration of PIV video cameras in Scheimpflug condition *Particle Image Velocimetry: Recent Improvements, Proc. of the EUROPIV 2 Workshop (Zaragoza, Spain, March/April 2003) Stanislas M (Berlin: Springer)* pp 391–405
[111] Louhichi H, Fournel T, Lavest J M and Aissia H B 2007 Self-calibration of Scheimpflug cameras: an easy protocol *Meas. Sci. Technol.* **18** 2616
[112] Cornic P, Illoul C, Cheminet A, Le Besnerais G, Champagnat F, Le Sant Y and Leclaire B 2016 Another look at volume self-calibration: calibration and self-calibration within a pinhole model of Scheimpflug cameras *Meas. Sci. Technol.* **27** 094004
[113] Astarita T 2012 A Scheimpflug camera model for stereoscopic and tomographic PIV *16th Int. Symp. on Applications of Laser Techniques to Fluid Mechanics (Lisbon, Portugal)*
[114] Sun C, Liu H, Jia M and Chen S 2018 Review of calibration methods for Scheimflug cameras *J. Sens.* **2018** 3901431

Dana Dabiri and Charles Pecora

Chapter 5

Particle tracking techniques

Once particle spatial locations are determined, the final step in PTV algorithms is tracking particles to determine the velocity field. There are many robust algorithms for matching particles from one frame to the next, which is necessary for properly calculating flow velocity. These algorithms can largely be used in both two- and three-dimensional problems once particle positions have been mapped to the spatial domain. Each section explains the extension of algorithms into the third dimension, which simply requires the addition of a z-coordinate to position vectors in many cases. In this section the following tracking algorithms will be discussed: multi-frame approaches for particle tracking using time-resolved PTV data (section 5.1); cross-correlation methods for tracking individual particles (section 5.2); relaxation methods for iteratively selecting probable matches based on the displacement of nearby particles (section 5.3); tracking algorithms based on various forms of neural networks (section 5.4); tracking that utilizes the velocity gradient tensor to identify strong deformations (section 5.5); tracking based on the similarity of nearby particle patterns using polar-coordinate similarities (section 5.6); optimization methods, which employ various iterative minimization techniques for determining particle tracks (section 5.7); tracking that finds matching using Delaunay tessellation and Voronoi diagrams to track particles (sections 5.8 and 5.9); and vision-based tracking methods, which match particles using the principles of proximity, similarity, and exclusion (section 5.10). Finally, techniques for identifying outlier particle tracks are discussed (section 5.11).

5.1 Multi-frame approach

Multi-frame tracking techniques are among the most common for PTV, as they use temporal information to identify a particle's trajectory over many time steps. The techniques use between two and four frames, which must be recorded at a regular time interval. This requirement for time-resolved imaging limits the applicability of these multi-frame techniques to either low-speed flows or high frame-rate cameras.

doi:10.1088/978-0-7503-2203-4ch5

Hassan and Canaan [1] introduced a particle tracking approach that works for singly exposed frames. The algorithm uses a four-frame sequence of time-resolved images to track a particle by determining the path that minimizes the variance in the particle's trajectory and velocity. Equations (5.1)–(5.5) calculate the total variance in a particle's path

$$\sigma_l = \sqrt{\frac{1}{3}[|d_{ij} - d_m|^2 + |d_{jk} - d_m|^2 + |d_{kl} - d_m|^2]}, \tag{5.1}$$

$$d_m = \frac{1}{3}(d_{ij} + d_{jk} + d_{kl}), \tag{5.2}$$

$$\sigma_\theta = \sqrt{\frac{1}{2}[|\theta_{ik} - \theta_m|^2 + |\theta_{jl} - \theta_m|^2]}, \tag{5.3}$$

$$\theta_m = \frac{1}{2}(\theta_{ik} + \theta_{jl}), \tag{5.4}$$

$$\sigma_t = \sqrt{\left[\frac{\sigma_l^2}{d_m^2} + \sigma_\theta^2\right]}. \tag{5.5}$$

d_{ij} is the distance between particle images i and j, potential matches between frames. θ_{ik} is the angle between the segments d_{ij} and d_{jk}, σ_l is the standard deviation of the track length, σ_θ is the standard deviation of the segment angles, and σ_t is the total standard deviation. This method makes no assumptions about the movement of neighboring particles, and does not exclude particles sharing a match in a frame.

Malik *et al* [3] introduced both the three-frame minimum acceleration (3MA) and four-frame minimum change in acceleration techniques (4MA) for 3D tracking. The 3MA technique uses a time sequence of three frames to determine the most likely path of a particle based on minimizing acceleration while the 4MA approach is based on minimizing the change in acceleration along a particle path. Both of these minimizations are performed based on triangulated 3D positions, although tracking can be similarly performed in 2D. These techniques require particle tracks for previous time steps, meaning that an initialization is required. It was suggested that this be done using a nearest neighbor heuristic (shown in figure 5.1(a)). A cost function, φ_{ij}^n, is defined as the distance between particle i in frame n and potential match j in frame $n + 1$:

$$\varphi_{ij}^n = \left\| x_j^{n+1} - x_i^n \right\|, \tag{5.6}$$

where the pair i, j corresponding to the minimum φ is selected as a match. This nearest neighbor criterion can be improved by using an initial estimate of local mean flow velocity either from known flow relations or nearby tracks. Once an initial track consisting of two points has been obtained, the 3MA criterion (shown in figure 5.1(b)) can be used. The acceleration is calculated for each three-point track

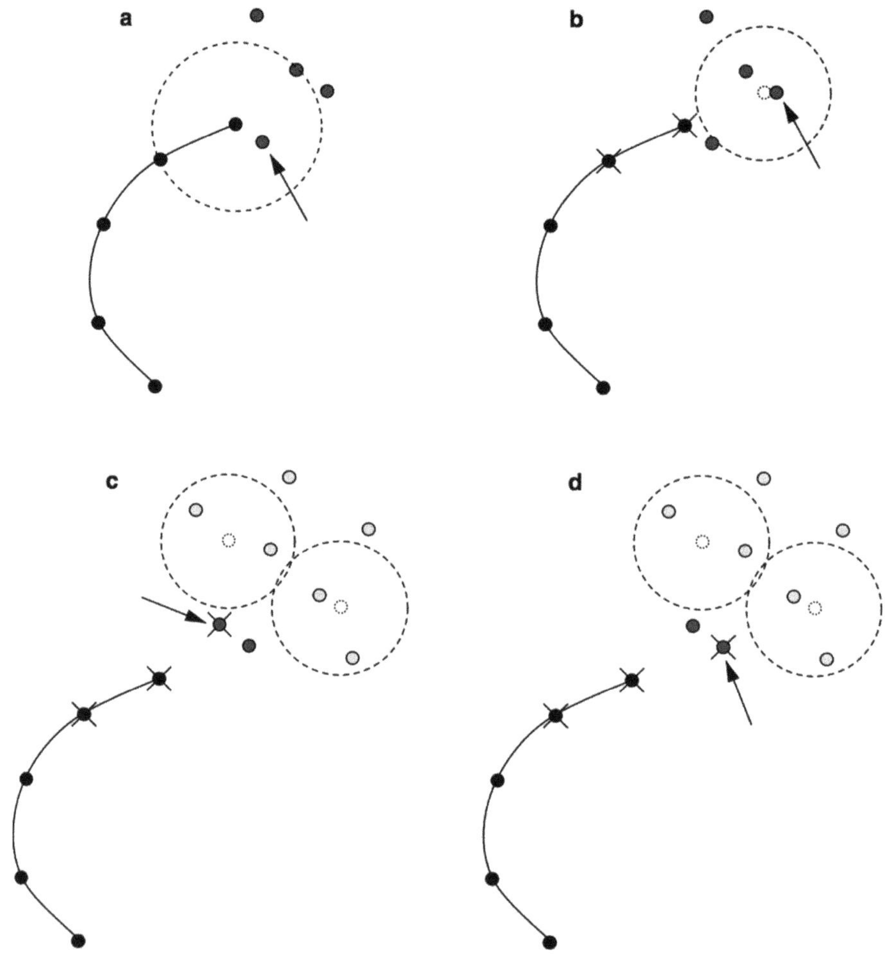

Figure 5.1. The heuristics used for multi-frame particle tracking algorithms: (a) nearest neighbor, (b) three-frame minimum acceleration, (c) four-frame minimum change in acceleration, and (d) four-frame best estimate. In each case the arrow points to the particle image that will be selected. Reprinted from [2] with permission of Springer.

$$\varphi_{ij}^{n} = \frac{\left\| x_j^{n+1} - 2x_i^n + x_i^{n-1} \right\|}{2\Delta t^2}, \tag{5.7}$$

where Δt is the time elapsed between frames. The 4MA algorithm (shown in figure 5.1(c)) expands upon this by minimizing the change in acceleration once a three-point track has been established

$$\varphi_{ij}^{n} = \frac{\left\| x_j^{n+2} - 2x_i^n + 1 + x_i^n \right\| - \left\| x_j^{n+1} - 2x_i^n + x_i^{n-1} \right\|}{2\Delta t^2}. \tag{5.8}$$

Ouellette *et al* [2] introduced a modification called the four-frame best estimate (4BE) approach (shown in figure 5.1(d)). Instead of minimizing the change in

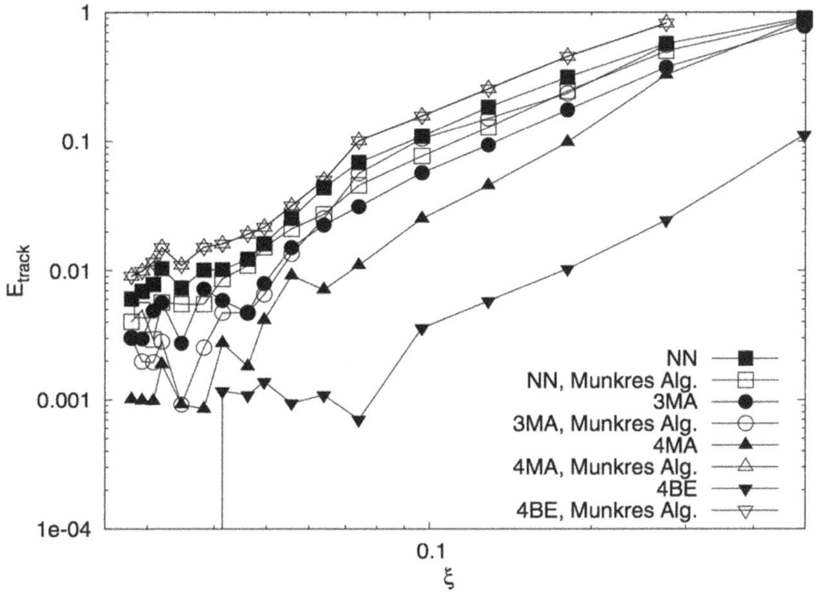

Figure 5.2. Comparison of the nearest neighbor (NN) method, three-frame minimum acceleration (3MA), four-frame minimum change in acceleration (4MA), and four-frame best estimate algorithm (4BE). Reprinted from [2] with permission of Springer.

acceleration, the difference between the projected fourth particle position, $\tilde{x}_i^{n+2}$, and the actual fourth particle position, x_j^{n+2} is minimized

$$\varphi_{ij}^{n} = \left\| x_j^{n+2} - \tilde{x}_i^{n+2} \right\|. \tag{5.9}$$

The projected position is based on the known positions and velocities at the first three points assuming constant acceleration. A conflict arises when a candidate particle is considered a best match for two reference particles in the most recent frame, requiring either starting a new track with the new particle, or using a cost minimization function. If the cost minimization approach is used, it is calculated using the Munkres algorithm. Figure 5.2 compares three- and four-frame algorithms using one of these two approaches. E_{track} is the ratio of the number of tracks that were imperfectly recreated to the total number of tracks and ξ is the ratio of average particle displacement to the average particle spacing. Figure 5.2 also shows that the 4BE algorithm outperforms the nearest neighbor (NN), three-frame minimum acceleration, and four-frame minimum change in acceleration techniques. Additionally, the Munkres algorithm degraded the performance of each algorithm except the nearest neighbor.

Tarlet *et al* [4] introduced a three-frame technique that minimizes acceleration for images of colored particles. In using color recognition software, the minimum acceleration approach is capable of identifying particle matches in increased seeding densities using color discrimination. It was found that the use of three different colored tracer particles effectively decreased ξ by 39%.

Although the particle density was not given for this comparison, the performance of these multi-frame algorithms depends on ξ, which would not only affect the average particle spacing, but also the average particle displacement through ξ, which according to figure 5.2 should be kept less than 0.2 if E_{track} is to be no larger than 0.01. For example, in a 2D application, for 0.03 N_{ppp}, the average particle spacing is ~5.8 pixels, suggesting that the average particle displacement should be no more than 1.2 pixels.

5.2 Cross correlation

A common technique for matching particles in two-frame PTV analysis uses the cross-correlation method [5–7]. It is based on the principle that a fluid particle and its closest neighbor particles typically move in quasi rigid-body motion. This knowledge is used to determine the probability of a candidate particle in the first frame being a match with a candidate particle in the second. Figure 5.3 shows a visual representation of this method.

In the cross-correlation algorithm proposed by Yamamoto *et al* [5] and Hassan *et al* [6], analysis was performed on a processed image pair, in which particle images have been identified and binarized in order to only use the particle positions rather than pixel intensities. This technique is called binary cross correlation (BCC), and has been implemented in 3D by placing synthetic spherical particles at the location of each triangulated particle location [5]. In the BCC method, a sub-region is defined around the reference particle, i, and each candidate particle, j, with N_i and N_j neighbor particles, respectively. The neighbor within the superimposed sub-regions that have overlapping spheres are considered. A parameter, S is defined as the ratio

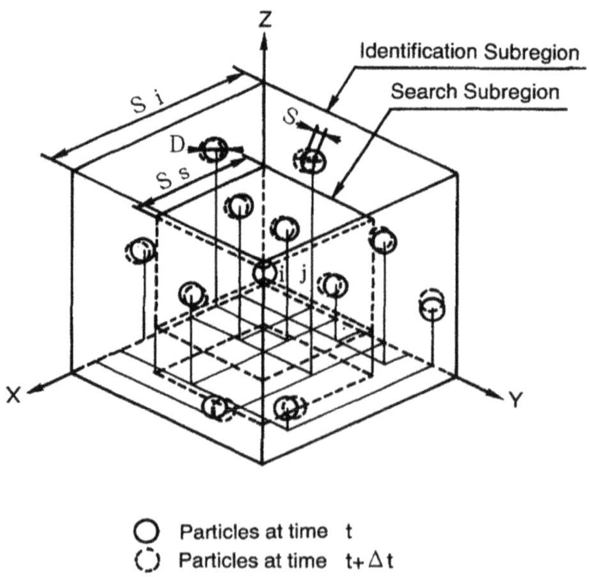

Figure 5.3. Particle positions from consecutive PTV frames with reference particle i overlapped with candidate particle j. Reproduced from [5], copyright The Japan Society of Mechanical Engineers.

norm of the 3D distance between the N overlapping particle centroids in the two sub-regions and the synthetic spherical particle diameter. Then the correlation coefficient for each candidate particle is calculated as

$$C_{ij} = \frac{\sum_{k=1}^{N} 1 - \frac{3}{2}S_k + 1/2S_k^3}{\sqrt{N_i N_j}}. \tag{5.10}$$

This correlation coefficient is equivalent to the calculation of the total overlapped volume between the particles in the reference and candidate sub-regions.

A different cross-correlation algorithm was proposed by Saga *et al* [7] and does not use the binary image data. Instead, the image gray values are used so as to allow CCD values to be used directly. The sub-regions search zones are defined similarly to the implementation by Yamamoto *et al* [5], but the correlation windows were varied in size in order to decrease computation time. Initially, correlation was performed using an 8×8 pixel window. The correlation coefficient, R, is calculated as

$$R = \frac{\sum_M \sum_N (A_{mn} - \bar{A})(B_{mn} - \bar{B})}{\sqrt{\left(\sum_M \sum_N (A_{mn} - \bar{A})^2\right)\left(\sum_M \sum_N (B_{mn} - \bar{B})^2\right)}}, \tag{5.11}$$

where A and B are the gray value matrices for the reference and second sub-regions, their mean values being $\bar{A}$ and $\bar{B}$. For candidate particles with a correlation coefficient above 0.6, a second correlation is performed using a 16×16 window. The threshold was then increased to 0.75. After another iteration or two, the candidate particle with the maximum correlation coefficient was selected. This cross-correlation algorithm was compared to the BCC algorithm using the VSJ standard image with 1997 identifiable synthetic particles (0.03 N_{ppp}) generated using LES simulation of an impinging jet flow. The results of this comparison are given in table 5.1.

The BCC method works for irrotational flows with small velocity gradients. The motions of rotational flows and flows with large velocity gradients prevent the cross-correlation algorithm from correctly recognizing neighbor particle motion.

5.3 Relaxation methods

Similar to the BCC method, the relaxation method matches a particle by comparing its motion to that of its neighbors. The difference, however, is that in the relaxation

Table 5.1. Comparison of BCC and cross-correlation algorithm proposed by Saga *et al* using VSJ standard image 01 [7].

Method	Particles tracked	Valid particle tracks	Valid N_{vpp}
BCC	917	881	0.013
Saga *et al* [7]	1592	1573	0.024

method, neighbor particles need not exactly overlap. This allows the algorithm to adjust for velocity gradients and rotation within the flow.

Relaxation methods were introduced by Rosenfeld *et al* [8] for labeling features in computer vision. The introduction of the technique to PTV was the result of work done by Wu and Pairman [9] followed by its application to turbulent flows by Baek and Lee [10]. Ohmi and Li [11] further modified the method of Baek and Lee, although Pereira *et al* [12] state that these modifications were proven to yield only minor improvements in their 3D implementation. Brevis *et al* [13] showed that combining cross correlation with relaxation could reduce computation time and improve results for inhomogeneous seeding densities. Jia *et al* [14] implemented a bidirectional computation of the relaxation method, which improved the handling of no-match particles.

The most important aspect of the relaxation algorithm is that the analysis is based on the particle match probability between the first and second frames, defined for every potential pair of particles inclusive of the probability of there being a loss-of-pair. Typically, a potential match is any candidate particle within a specified maximum displacement distance of the reference particle. The particle match probabilities are updated iteratively based on the match probabilities and relative motions of neighboring particles. The no-match probability accounts for particles that leave the observation volume from one frame to the next. As relaxation algorithms are iterative, the match probabilities must be initialized. Figure 5.4 shows the initialization method and its formulation is

$$P_{ij(i)}^{(0)} = P_i^{*(0)} = \frac{1}{n_i + 1}, \tag{5.12}$$

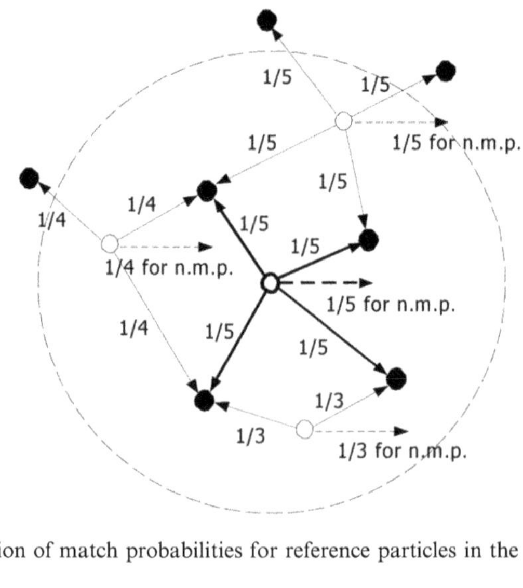

Figure 5.4. Initialization of match probabilities for reference particles in the relaxation method [11].

where P_{ij} is the probability that the reference particle i is a match with candidate particle j. P_i^* is the no-match probability of particle i and n_i is the number of candidate particles within the maximum displacement distance of particle i. Match probabilities can alternatively be initialized using results from prior computations, such as cross correlation. This typically shortens the computation time of the relaxation method.

After initializing the match probabilities, the match probabilities are updated for each iteration, n. In order to update the match probability between a reference particle, j, and a candidate particle, i, the contribution from the match probability of each neighboring particle, k, and its candidate particle, l, is

$$\tilde{P}_{ij}^{(n)} = P_{ij}^{(n-1)}\left(A + B\sum_k \sum_l P_{kl}^{(n-1)}\right), \tag{5.13}$$

where A and B are constants in the equations, set to be 0.3 and 3.0, respectively, as suggested by Barnard and Thompson [15]. The superscripts signify that the probabilities from iteration $n - 1$ are used to generate the probabilities for iteration n. The contribution of neighboring particles in the double summation above is limited to the neighboring matches that that satisfy

$$|d_{ij} - d_{kl}| < R_c, \tag{5.14}$$

where d_{ij} is the displacement vector from particle i to particle j and d_{kl} is that from particle k to particle l. A threshold, R_c, is the radius of the relaxation area, which allows some deviation from parallel motion. The updated probabilities are then normalized using

$$P_{ij}^{(n)} = \frac{\tilde{P}_{ij}^{(n)}}{\sum_j \tilde{P}_{ij}^{(n)} + P_i^{*(n-1)}}, \tag{5.15}$$

which uses the no-match probability from the preceding iteration. The no-match probability is not updated, but just renormalized as

$$P_{ij}^{*(n)} = \frac{\tilde{P}_{ij}^{*(n-1)}}{\sum_j \tilde{P}_{ij}^{(n)} + P_i^{*(n-1)}}. \tag{5.16}$$

This update scheme is iterated 10–20 times, ideally resulting in correct match probabilities near unity and others near zero. A threshold for the maximum difference between iterations is defined to determine when the relaxation algorithm has converged. The highest match probability (or no-match probability) is then selected as the correct track for each candidate particle.

Ohmi and Li [11] suggested two changes to this algorithm. It was pointed out that the update procedure for the no-match probability lacked consistency and could only increase if all other matching probabilities thoroughly decreased, which is rare in densely seeded flows. In order to overcome this, a new scheme for updating the no-match probability was suggested as

$$\tilde{P}_i^{*(n)} = \sum\nolimits_{m<C} \frac{DP_i^{*(n-1)}}{n_i}, \tag{5.17}$$

where C and D are constants, equal to 0.1 and 5.0, respectively, and m is the ratio of the number of interparticle links satisfying equation (5.17). To the number of candidate particles from all the neighboring particles. This causes the no-match probability to increase at a constant rate if too few neighbor particles appear to undergo quasi-parallel motion. After this update, the normalization step needs to be modified such that

$$P_{ij}^{(n)} = \frac{\tilde{P}_{ij}^{(n)}}{\sum_j \tilde{P}_{ij}^{(n)} + P_i^{*(n)}}, \tag{5.18}$$

and

$$P_{ij}^{*(n)} = \frac{\tilde{P}_{ij}^{*(n)}}{\sum_j \tilde{P}_{ij}^{(n)} + P_i^{*(n)}}. \tag{5.19}$$

The second change that the threshold for R_c should vary with local flow velocity. Ohmi and Li [11] suggested that low-speed flow regions will produce clusters of quasi-parallel but clearly erroneous vectors if the same threshold is used. Thus, this local threshold is

$$R_c = E + F|x_i - y_j|, \tag{5.20}$$

where E and F are constants and y_j is the position of the candidate particle. E is image dependent and typically in the range of 1.0–4.0. F is a proportionality constant that can be fixed to 0.05. Despite the seemingly small contribution of this proportionality constant, it reduces the number of spurious vectors in strong shear flows.

To assess the performance of this algorithm, tests were performed using the VSJ PIV standard image 301, which has 4000 synthetic particle images with 0.06 N_{ppp} simulating the vertical portion of an impinging jet flow. It should be noted that not all of the 4000 particle images were identifiable, and the number of identified particle images was not reported. Table 5.2 gives the results of the comparison of three tracking techniques. It can be seen that the relaxation method outperforms the BCC method in particle matching. Also, the improvements made by Ohmi and Li [11] provided an improvement over the original relaxation algorithm both in number of particle tracks obtained and in reliability. The resulting density of valid vectors obtained using the improved relaxation method was 0.012 vectors per particle (N_{vpp}).

Table 5.2. Comparison of four particle tracking algorithms using the PIV standard image 301 [11].

Algorithm	Number of particles matched	Correct matching rate (%)
Four frame	779	86%
Binary cross correlation	860	91%
Original relaxation	786	96%
Ohmi and Li relaxation	808	98%

Brevis *et al* [13] introduced an algorithm that combines the strengths of cross correlation (section 5.2) and the relaxation methods. Figure 5.5 outlines how the two techniques are combined. Matches are initially found using cross correlation, which has fast computation time and works well for low particle density, low gradient, irrotational flows. A cross-correlation threshold, double match filter, mean filter, and median filter (described in detail in section 5.11) are then used to identify correct tracks from the cross-correlation analysis. Tracks that pass the filter are assigned a match probability of one, essentially removing them from relaxation calculations. The remaining tracks then use the cross-correlation coefficients, renormalized using a no-match probability equal to the difference between one and the maximum correlation coefficient, as initial match probabilities for the relaxation algorithm. For these particles and regions in which cross-correlation fails to identify a match, the relaxation method matches the remaining particles. A double match filter and median filter are then applied to remove spurious vectors from the final results. Additionally, as the relaxation method starts with estimates for match probabilities, fewer iterations are required for convergence.

Brevis *et al* [13] compared their hybrid algorithm to the cross-correlation and relaxation methods using exact particle locations from VSJ standard image 301, which has 4042 possible vectors (0.062 N_{vpp}) simulating a jet impingement on a wall. The hybrid algorithm achieved 98.46% yield (equation (3.15)), which was roughly 5% greater than the cross-correlation method and 1% greater than the relaxation method. Although the hybrid algorithm performed only marginally better than the relaxation method, it was able to achieve its results several times faster as a result of the computational efficiency of the cross-correlation algorithm. Additionally, the synthetic flow data had homogeneous particle density, meaning that there were no regions in which particles were scarce enough to give the cross-correlation method an advantage. The algorithm was also applied to experimental time-resolved data at a seeding density of 0.0045 N_{ppp}. At this low density, particle tracks longer than 100 frames were extracted and are shown in figure 5.6.

Further improvements to the relaxation method were made by Jia *et al* [16] using Delaunay tessellations. The range for neighbor particles included in the relaxation parameter calculation is determined automatically rather than set to a range. Only particles with direct links to the target particle within the Delaunay tessellation are considered. Additionally, a method of bidirectional computation improves the handling of no-match probability. First, no-match probability is directly ignored

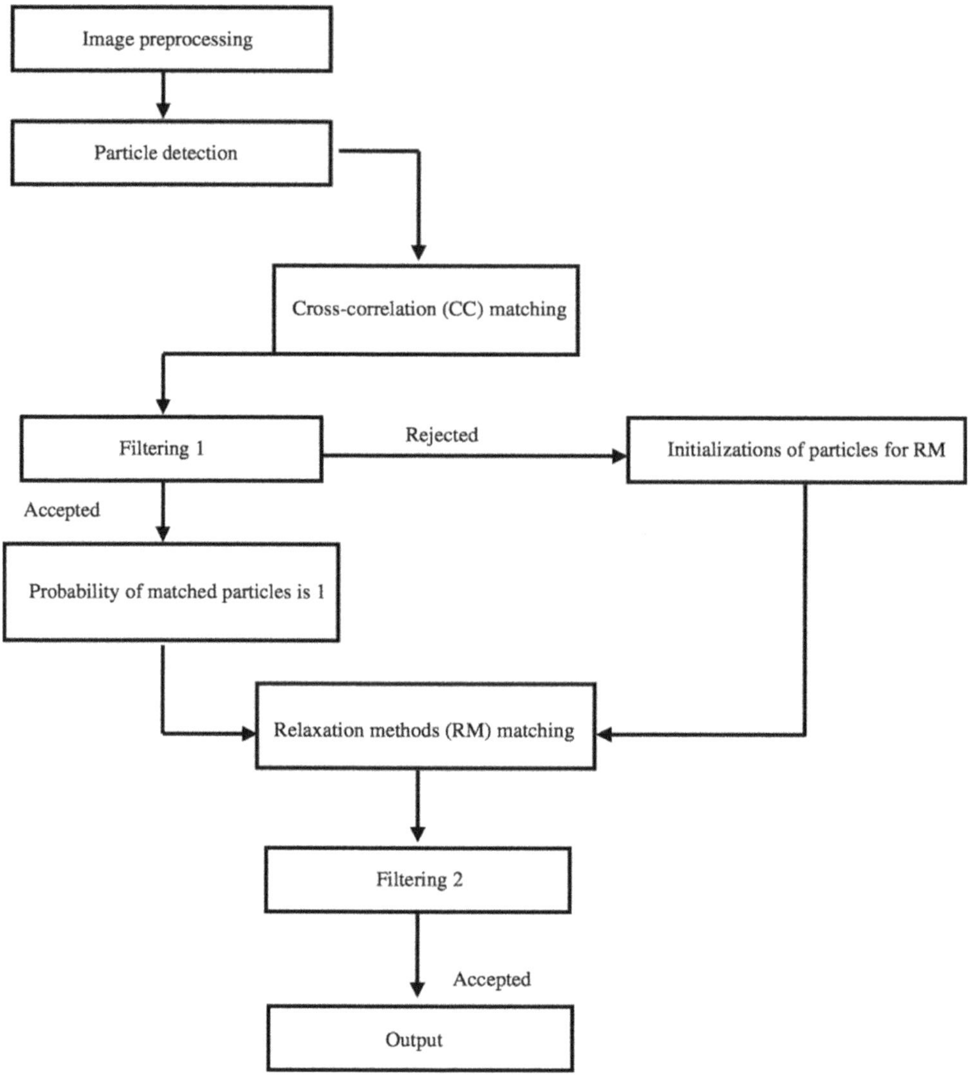

Figure 5.5. Flow chart for the Brevis *et al* hybrid algorithm. Reprinted from [13] with permission of Springer.

for calculation. Second, calculations are done for the match probabilities, P_{ij}, both from the first frame to the second and from the second frame to the first. Then the latter calculation is converted and merged with the former. Considering the particle tracks achieved from both forward and backward computations, final results are selected based on if these tracks that do not conflict with one another. When conflicts arise, tracks that appear more often are selected, and ones that appear less often are discarded. If two tracks conflict and appear the same number of times, they are regarded as no-match particles.

Results from the bidirectional algorithm are compared to those of the relaxation scheme of Ohmi and Li [11] at 0.046 N_{ppp} in figure 5.7, where η is the fraction of

Figure 5.6. Long particle tracks obtained using 3D implementation of the ICCRM algorithm. Reprinted from [13] with permission of Springer.

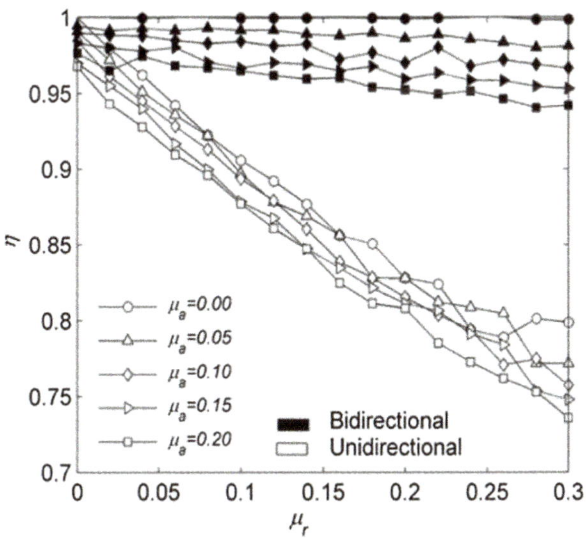

Figure 5.7. Comparison of the unidirectional and bidirectional algorithms in the presence of removed and added particles [16].

correctly matched vectors, μ_a is the ratio of randomly added particles to total particles, and μ_r is the ratio of removed particles to total particles. The bidirectional algorithm showed improved performance in the case of removed particles. The computation time of the bidirectional scheme is twice that of the unidirectional one, however, the cost was considered acceptable due to the improvement in performance. Note that for the case of zero removed particles, the bidirectional algorithm achieved 0.046 N_{vpp}, as all possible vectors were identified correctly.

Jia *et al* [14] also performed a 3D test of the bidirectional relaxation algorithm using four different VSJ 256 × 256 pixel standard images. The parameters of each are detailed in table 5.3. The maximum particle densities of images 301C, 337, 351, and 377 are 0.064, 0.105, 0.033, and 0.015 N_{ppp}, respectively. The ratio of correctly identified tracks to available tracks, η, is plotted as a function of the ratio of erased particles to the total number of particles, μ. For all images and ranges of μ, the bidirectional algorithm performs substantially better than the unidirectional one. The highest resulting vector densities using the bidirectional algorithm for images 301C, 337, 351, and 377 are 0.058, 0.095, 0.031, and 0.015 N_{vpp}, respectively. The yield decreases as the ratio of erased particles increases for all seeding densities (figure 5.8).

In an improved technique called enhanced particle tracking velocimetry (EPTV) [17], performance was improved by including a match probability initiation based on the diameter of the particle image. The initial match probability for reference *i* and candidate *j* is

$$P_{ij} = \alpha\left(1 - \frac{|d_i - d_j|}{\Delta d_{\max}}\right) + (\alpha - 1)\left(1 - \frac{|E_i - E_j|}{\Delta E_{\max}}\right), \tag{5.21}$$

where

$$E_i = \sum\nolimits_{k,l \in S} I_{k,l}, \tag{5.22}$$

and

$$d_i = \sqrt{\frac{8E}{\pi I_0}}. \tag{5.23}$$

The values E_i and d_i are the total image intensity and image diameter, respectively. A particle image segment area, S, contains pixel intensities, $I_{k,l}$, with a peak value, I_0. The weight coefficient, α can vary from 0 to 1 based on the set-up parameters. For instance, if the particle diameter is less than one pixel, α is set to zero so that the match probability is solely determined by the total image intensity.

The results for synthetic images with 0.062 N_{ppp} are compared in table 5.4 for EPTV and other algorithms. Not only was the correct match rate improved over the variational approach, discussed in section 5.7.2, and previous relaxation method, but the number of found vectors more than doubled to 0.026 N_{vpp}. The vast difference in yield between EPTV and the other algorithms is due to the different particle identification algorithms used. The dynamic threshold binarization algorithm used by Mikheev and Zubtov identified twice as many particles as the simple threshold binarization of Ohmi and Li and the particle mask correlation of Ruhnau *et al* [18]. Thus, it is difficult to directly compare the performance of the tracking algorithms.

Table 5.3. VSJ 3D standard images used for comparison of unidirectional and bidirectional relaxation algorithms [17].

Items	Serial number of flows			
	301C	337	351	377
Flow description	Slit light sheet	Stereo-PIV images	3 angles, wall refraction	3 angles, unknown wall refraction
Frame number	145	201	145	201
Max particle number	4202	6883	2138	979
Min particle number	3737	6509	1950	899
Max inter-frame particle difference	54	51	29	13
Min inter-frame particle difference	−40	−44	−31	−14
Max particle entering ratio (%)	5.61	**3.13**	3.49	2.64
Min particle entering ratio (%)	2.7	1.58	1.55	0.21
Max particle escaping ratio (%)	5.82	3.39	3.65	2.08
Min particle escaping ratio (%)	2.72	1.49	1.57	0.11

Inter-frame particle difference is defined as $N_{P1}–N_{P2}$; N_{P1} and N_{P2} are particle number in frame t and frame $t + \Delta t$, respectively. The entering and escaping ratios are based on N_{P1} in a match.

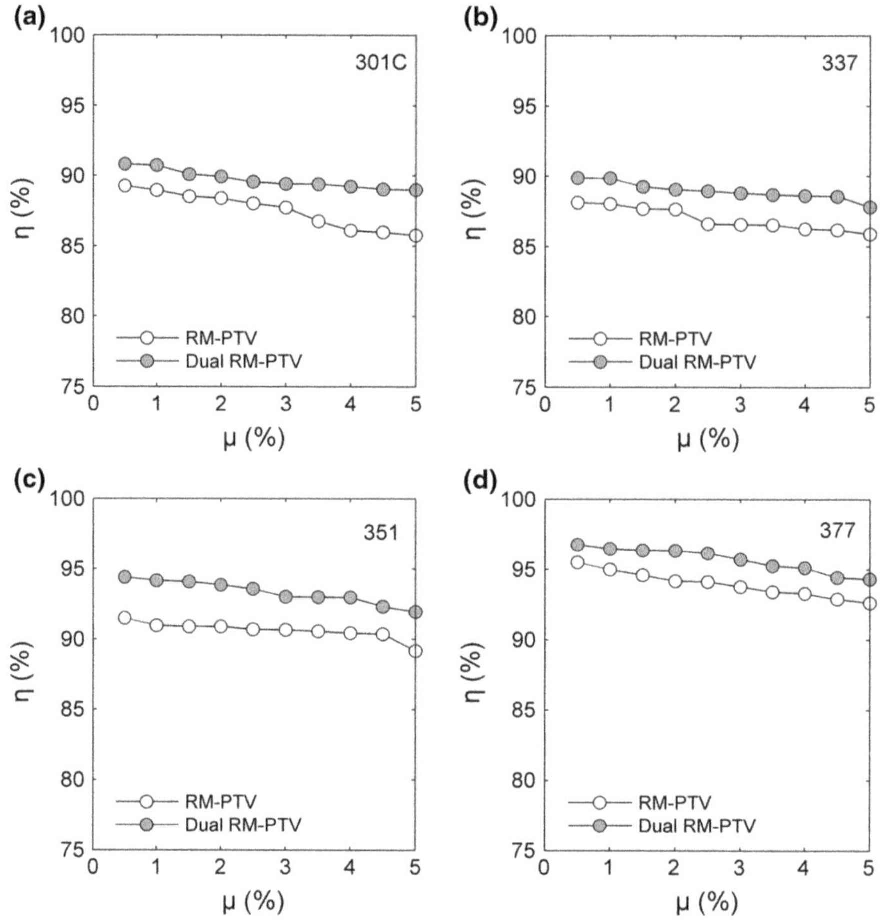

Figure 5.8. Comparison of bidirectional (dual RM-PTV) and unidirectional (RM-PTV) algorithms for four VSJ 3D standard images. Reprinted from [14] with permission of Springer.

5.4 Neural networks

A PTV algorithm based on neural networks was introduced by Grant and Pan [19]. Neural networks are commonly used in data science and are modeled based on the structure of the human brain, which is thought to consist of a three-dimensional matrix of interconnected neurons. This structure is capable of implementing simultaneous nonlinear processing strategies. Information is input to neurons, which modify and redistribute the information. Inputs into each neuron are summed, and if the net input exceeds a threshold, then the neuron is activated and transmits a signal to other neurons (shown in figure 5.9, left). Weights or interconnection efficiencies determine the interaction of information between neurons. In both the human brain and in neural networks these weights can update to learn from input datasets. This characteristic of neural networks allows a system to learn from examples and adapt. The performance of a neural network largely depends on the availability of relevant

Table 5.4. Comparison of EPTV on a VSJ standard image #301 [17].

PTV method	Frame pair	Number of vectors found (N_F)	Number of correct vectors (N_c)	Correct matching rate (%) N_c/N_F
True data	0–1	4042	4042	100
	0–2	3917	3917	100
EPTV	0–1	1759	1733	98.6
	0–2	1521	1505	98.9
NRX (Ohmi and Li 2000)	0–1	808	788	98
	0–2	714	680	95
YAR (Ruhnau *et al* 2005)	0–1	872	865	99
	0–2	904	885	98

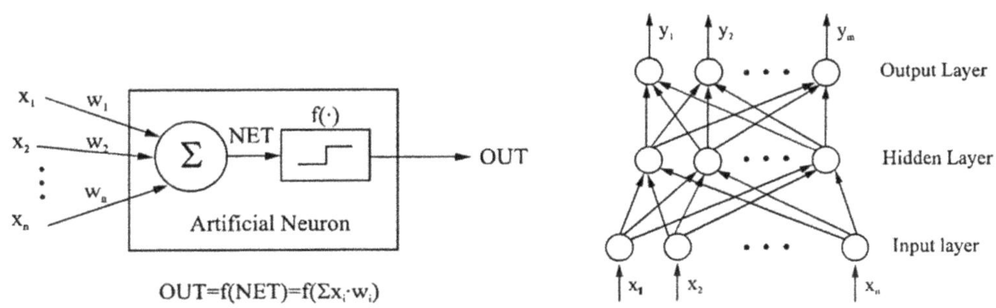

Figure 5.9. Inputs and activation of an artificial neuron (left) and structure of neurons in a feed-forward network (right). Reprinted from [19] with permission of Springer.

examples. Neuron activations and weights continue to interact through layers until an output layer is reached (shown in figure 5.9, right). The result from a neural network algorithm is an output label or classification, which is determined by identifying the neuron within the output layer with the greatest activation.

The neural network for particle tracking suggested by Grant and Pan (described by Labonté [20]) is a feed-forward network with three layers of neurons. The three layers are all the same size, square with $2N_{\max} + 1$ nodes per side, where $N_{\max}$ is the integer pixel value of the maximum displacement of a particle. A binary matrix is made for each of two consecutive PTV images, in which each element corresponds to a pixel. The value of a pixel is set to one if a particle image resides in the pixel, and it is otherwise equal to zero. The input into the neural network is a matrix, R, of the same size as the neuron layers. There is a single input matrix for each reference particle in the first PTV frame. The input matrix consists of a square sub-region of the binary matrix from the second PTV image centered at the coordinates of the reference particle. The first layer of the neural network takes the input matrix, R_{mn}, and weights, W_{mn}, to generate an output matrix

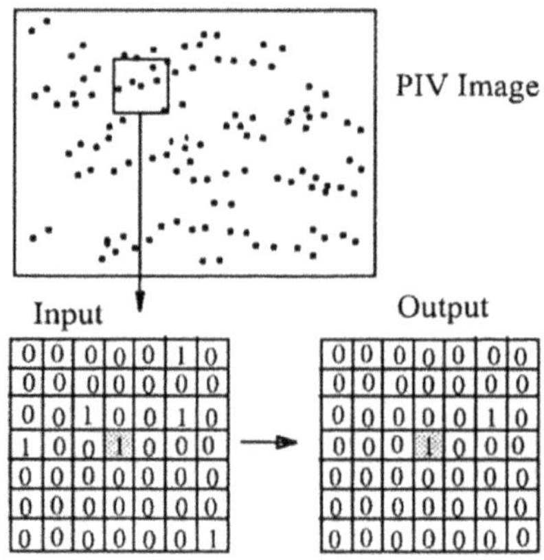

Figure 5.10. Example input and output of the neural network used by Grant and Pan. Reprinted from [19] with permission of Springer.

$$S_{mn} = \mathrm{H}(R_{mn}W_{mn} - T_1), \tag{5.24}$$

where T_1 is a threshold and 'H' is the Heaviside function. The output S is then used as the input to the second layer of the neural network. The second layer outputs the product of S_{mn} and the neuron weights U_{mn}. The neuron with the maximum activation, which must also exceed a set threshold, wins and is set to one while the other neurons are set to zero. The winning neuron identifies the candidate particle that is part of the reference particle's track. This is illustrated in figure 5.10.

The neuron weights are only updated when the winner is identified with a one. If the winner's coordinates are (m^*, n^*), I_{mn} is the Euclidean distance between the winner and the reference particle, $\rho_{mn} = (N_{\max} - I_{mn})/N_{\max}$, and α is a constant between zero and one, then the weights are updated as

$$W_{mn} = \mathrm{Max}\{\rho_{mn}, W_{mn}\} \text{ for the first layer, and} \tag{5.25}$$

$$U_{mn} = U_{mn} + \alpha(\rho_{mn} - U_{mn}) \text{ for the second layer,} \tag{5.26}$$

which is then normalized as

$$U_{mn} = U_{mn}/U_{\max}, \tag{5.27}$$

where $U_{\max}$ is the maximum weight in the second layer of the neural network. It was suggested that the inputs for spatially adjacent particles should be updated successively, for instance the candidate particles in a $D_{\max} \times D_{\max}$ pixel square patch in the lower right-hand corner were processed first. Then each square patch along the bottom was processed in a row. Then the next row of patches above that was processed, and so on. Weight initialization for this method requires statistical estimates of local flow conditions or a learning set of data that is correctly labeled.

A paper by Labonté [20] gives a different technique for applying neural networks to particle tracking. The method creates a self-organizing map (SOM) that constantly shifts particles based on the neighboring particle displacements. This is done to approximately reconstruct the flow of the fluid from the two pictures of tracer particles. Two sub-networks are created, each with one layer of neurons. Each neuron represents a particle image, thus the two subnets represent the two PTV images. The neurons have a value of $\boldsymbol{x}_i$ for particles in the first image and $\boldsymbol{X}_i$ for particles in the second. Corresponding weights are $\boldsymbol{w}_i$ for neurons in the first sub-network and $\boldsymbol{W}_i$ for neurons in the second sub-network. These weights are initialized as the position vector of the associated particle image.

In order to shift the particles such that they overlap, the weights between layers are compared. For neuron j in the first sub-network, the weights of all neurons in the second sub-network are compared to compute the Euclidian distance between the two coordinates

$$d_{ij} = \| W_i - w_j \|, \tag{5.28}$$

where the neuron i corresponding to the minimum d_{ij} activates neuron j if d_{ij} is less than the maximum allowed displacement. Otherwise the neuron is not activated. The winning neuron weight, $\boldsymbol{W}_i$, and the candidate neuron weight, $\boldsymbol{w}_c$, are then used to update the weight of all j neurons in the first sub-network as

$$\Delta w_j(W_i) = \alpha_{jc}(W_i - w_c), \tag{5.29}$$

with

$$\alpha_{jc} = \begin{cases} \alpha & \mathit{if}\ |w_c - w_j| \leqslant r \\ 0 & \mathit{if}\ |w_c - w_j| > r \end{cases}, \tag{5.30}$$

where α is a constant between zero and one and r is a distance that defines which neighbor particles translate with the candidate. This update procedure influences particle tracks to move coherently with neighboring particle tracks.

Calculations are all performed independently for each neuron, then weights are all updated simultaneously for the first sub-network. The same procedure is then used to update the weights in the second sub-network. After this is completed, the neighborhood radius, r, is decreased and the weights are all updated again. This causes smaller and smaller groups of particles to move together with each iteration. The radius is decreased linearly until the two sets of weight vectors can be considered equal. The weight vectors can be considered equal if less than a user-defined tolerance. A larger tolerance allows for faster computation time (figure 5.11).

Unlike Grant and Pan's neural network method, the self-organizing map requires no initialization or training. Both algorithms were tested with a point vortex flow with 200 images per picture, 40 of which were unmatchable. The ratio of mean particle displacement to particle spacing, ξ, was 0.26. The image resolution was not given, so no particle density was reported. Grant and Pan's algorithm correctly matched 55% of the particles, while Labonté's algorithm achieved 78% success (figure 5.12).

Ohmi [21] suggested a modification to this self-organizing map that involves a more appropriate learning coefficient, α_{jc}, in equation (5.31), which is based on a distance-dependent Gaussian function. The learning coefficient is often the same as

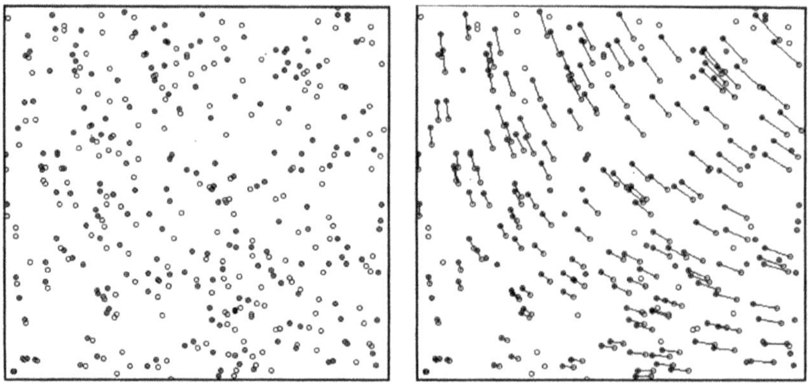

Figure 5.11. Visualization of a self-organizing map, in which particles from two frames are overlaid and organized until a match is found for each. Reprinted from [20] with permission of Springer.

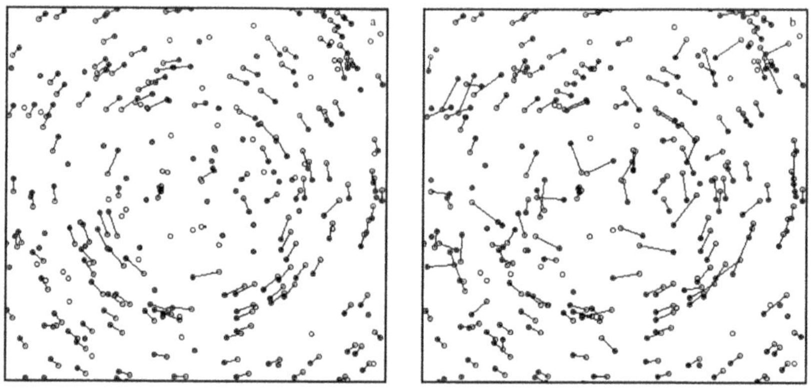

Figure 5.12. Displacement fields determined by the neural nets of Labonte (left) and Grant and Pan (right). Reprinted from [20] with permission of Springer.

before, but now neighboring particles are updated based on a continuous distance function rather than in a binary manner as in equation (5.30):

$$\alpha_{jc} = \begin{cases} \alpha & \text{if } |w_c - w_j| \leqslant r \\ \alpha e^{-(|w_c - w_j| - r)^2/2r^2} & \text{if } |w_c - w_j| > r \end{cases}. \tag{5.31}$$

A second modification deals with loss-of-pair particles which was borrowed from a solution to the traveling salesman problem (TSP) algorithm [22, 23]. In their algorithm, the output layer neurons can be removed or doubled according to their computational history. In the context of this algorithm, a neuron that is not identified as a winner for ten consecutive iterations is removed. Output neurons that are identified as winners for multiple input neurons in three consecutive iterations are duplicated and treated as independent neurons for the remainder of the algorithm. This aims to improve convergence of the algorithm in the presence of loss-of-pair particles. The improvement of these changes is shown in figure 5.13

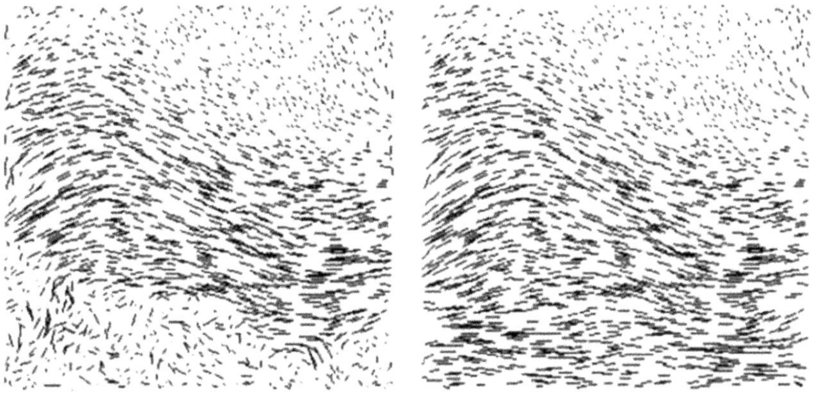

Figure 5.13. Comparison of the original SOM algorithm (left) and the modified SOM (right) for PIV VSJ standard image #01. Reprinted from [21], Copyright (2008), with permission from Elsevier.

using the VSJ standard image #01 with 0.033 N_{ppp}, where the new algorithm has many fewer spurious vectors than the original. It should be pointed out that the author did not present quantitative measurements of the algorithm's improvement over the neural network of Labonté.

The self-organizing map algorithms use particle position vectors as inputs, thus can be extended into three dimensions. Ohmi [21] extended his algorithm to work to 3D in this manner and applied it to VSJ standard image #351 to track 2500 particles, or 0.038 N_{vpp}; however, quantitative results for the yield and reliability were not reported.

5.5 Velocity gradient tensor

Cross-correlation methods determine particle displacements best when particles move as non-rotating rigid bodies. As such, these techniques cannot always perform well when strong velocity gradients occur within a flow. The velocity gradient tensor (VGT) tracking method accounts for translation, rotation, and strain when determining particle tracks [24–26].

Like other methods, the VGT technique chooses potential match particles J in the second frame for each particle I in the first frame. In doing so, a velocity $\boldsymbol{u}(\boldsymbol{x}_I)$ is computed for a time step Δt

$$\boldsymbol{x}_J = \boldsymbol{x}_I + \boldsymbol{u}(\boldsymbol{x}_I)\Delta t. \tag{5.32}$$

In the cross-correlation method, it is assumed that the neighbors of particle I, x_{Ik}, translate with the same velocity to neighbors of particle J, x_{Jk}. The VGT method, however, accounts for differences in velocity such that

$$\boldsymbol{x}_{jk} = \boldsymbol{x}_{ik} + \boldsymbol{u}(\boldsymbol{x}_{ik})\Delta t. \tag{5.33}$$

The velocity of each neighboring particle can be approximated using a first-order Taylor expansion around $\boldsymbol{x}_I$

$$x_{jk} = x_J + (I + \delta u(x_I)\Delta t)(x_{ik} - x_I), \tag{5.34}$$

where I is the identity matrix and δu is the velocity gradient tensor. A transformation matrix, A is then defined for simplicity. The velocity gradient tensor is incorporated as

$$A = I + \delta u(x_I)\Delta t. \tag{5.35}$$

For the local transformation in the neighborhood of particle I. The transformation is implemented for each of the n neighbors and a least squares minimization is performed as

$$E_{IJ} = \sum\nolimits_{k=1}^{n} |X_{J,k} - AX_{I,k}|^2, \tag{5.36}$$

where

$$X_{J,k} = x_{jk} - x_j. \tag{5.37}$$

Least squares minimization is performed for each combination and permutation of neighbors for particle I and J. The minimization is also performed for each possible candidate particle J. The particle pair (I,J) with minimum corresponding E_{IJ} is then used to finalize and compute a velocity vector. This procedure is outlined in the flow diagram in figure 5.14.

To test this algorithm in comparison to others, a Karman vortex-shedding flow was simulated numerically with the 2D incompressible Navier–Stokes equation and a Reynolds number equal to 500 [26]. The VGT algorithm used eight neighbor particles for least squares minimizations. The displacement ratio, Φ, denotes the ratio between the maximum particle displacement and the average distance between particles. The algorithms are compared for a range of displacement ratios and particle numbers in figure 5.15. All of the tested algorithms performed well for displacement ratios below 2, and the performance of each began to degrade as displacement ratio increased further. The performance of VGT tracking was superior to both the 4PTV and BICC methods for high displacement ratios. The authors pointed out that both the 4PTV and BICC methods failed to correctly track particles at the highly rotational vortex cores. No pixel resolution was reported, thus no particle or vector concentrations could be ascertained. The results show that the tested tracking algorithms performed better at higher particle densities for a given displacement ratio.

5.6 Polar-coordinate similarity

Ruan and Zhao [27] introduced a novel technique in 2005 that relates the similarities in polar-coordinate systems surrounding candidate particles for matching. Shindler *et al* [28] improved this polar-coordinate system similarity (PCSS) method by including spatial adaptivity and temporal extrapolation. In order to implement this method, particle centroid locations must be known in either two or three dimensions, although the algorithm has not yet been applied to 3D data. Figure 5.16

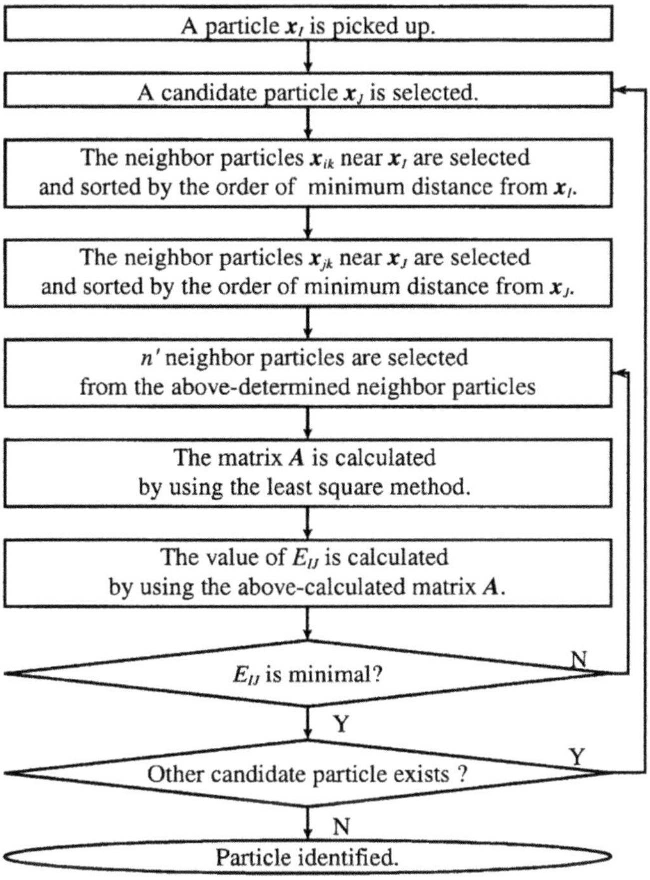

Figure 5.14. Flow chart for the pairing method. Reprinted from [25] with permission of Springer.

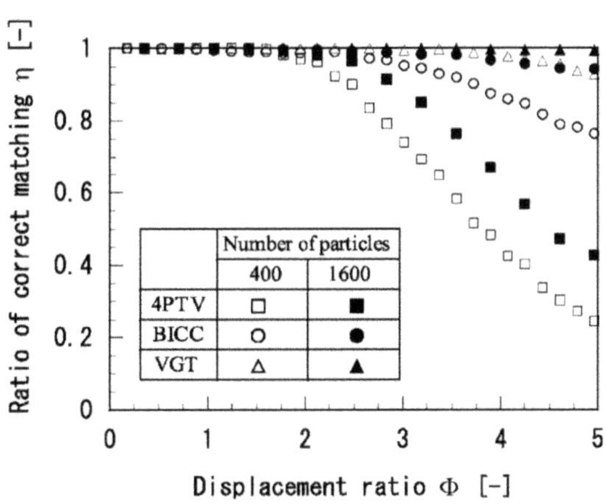

Figure 5.15. Performance comparison of four-frame minimum acceleration (4PTV), binary image cross correlation (BICC), and velocity gradient tensor (VGT) tracking methods for Karman vortex-shedding flow [26].

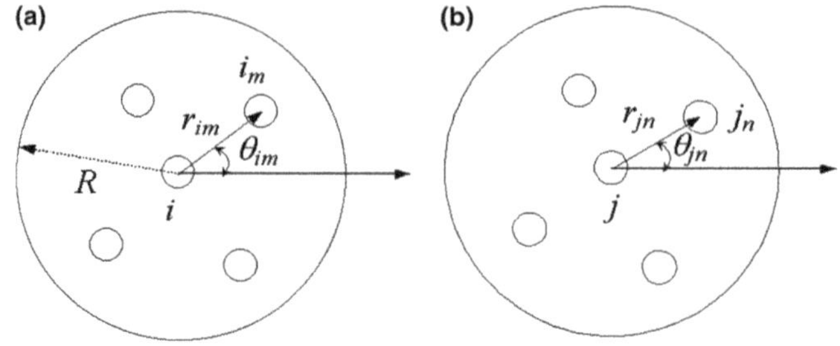

Figure 5.16. Concept of polar-coordinate similarity between particles *i* and *j*. Reprinted from [27] with permission of Springer.

Table 5.5. Comparison of the number of calculated vectors using the polar-coordinate similarity method (PCSS), cross correlation (BICC), and four-frame in-line tracking (4PTV) [27].

Method	Vectors	Remark
PCSS	664	after delete spurious vectors
BICC	650	
4PTV	604	need long computing time

shows how polar coordinates for surrounding particles are calculated with respect to reference particle i in the left frame and candidate particle j in the right.

The relative polar coordinates of neighbors to particles i and j are compared and summed to generate a similarity coefficient

$$S_{ij} = \sum_{n=1}^{N} \sum_{m=1}^{M} \mathrm{H}(\varepsilon_r - |r_{im} - r_{jn}|, \varepsilon_\theta - |\theta_{im} - \theta_{jn}|), \tag{5.38}$$

where $\mathrm{H}(x, y)$ is the Heaviside function and ε is the error allowed in radius coordinate and angle coordinate. In their implementation, the authors used $\varepsilon_r < 5$ pixels and $\varepsilon_\theta < 30°$. The candidate particle with the maximum similarity coefficient is selected as the match for each reference particle.

Table 5.5 compares this polar-coordinate similarity (PCSS) tracking algorithm with others. The test was done using two frames of experimental data of flow induced by an oscillating plate with 0.0026 N_{ppp}. The PCSS algorithm yielded 0.0025 valid N_{vpp}. The number of vectors reported for BICC and 4PTV algorithms included spurious vectors, thereby showing that the PCSS method outperformed the BICC and 4PTV methods.

The most substantial change made to the algorithm by Shindler *et al* [28] was the adaptive interrogation window size. Originally, a fixed radius, R, was used to find neighboring particles; however, for flows with regions of sparsely seeded flow, there could be insufficient neighbors to generate a similarity coefficient. Thus, it was suggested that a fixed number of nearest neighbors should be selected. Temporal extrapolation was also suggested for multi-frame tracking with the PCSS algorithm. If data is time-resolved and a particular reference particle in frame k has a track from the previous frame, $k - 1$, then a linear extrapolation can be performed to predict the particle's position in the next frame, $k + 1$. If this is done, then the search radius and number of candidate particles can be reduced. If a candidate particle's track spans from $k - 2$ to k, then extrapolation can be performed by assuming uniform acceleration. This can further reduce the search radius and number of candidate particles, which will result in lower computation time and increased probability of determining the correct track. An additional frame-gap technique was proposed, in which if a candidate particle in frame k does not find a match in frame $k + 1$, then a search is performed in frame $k + 2$ with double the search radius and a lower threshold for the similarity coefficient.

This algorithm was named enhanced polar-coordinate similarity, and in its two-frame implementation was called 2F-EPS. In the context of time-resolved data, a multi-frame enhanced polar-coordinate similarity (MF-EPS) algorithm used the velocity measurements from the first two frames to extrapolate a particle's future position. Table 5.6 shows both the performance enhancement of polar-coordinate similarity algorithms over three-frame and relaxation approaches and the benefits realized from EPS and the multi-frame extension. The test was performed using VSJ standard image 301, and the particle centroids were determined using the optical flow equation and a 1D Gaussian estimator (see section 1). The most substantial result is the improvement of MF-EPS using three images. Adding the third frame resulted in nearly triple the number of vectors, which is equivalent to 47% more vectors per time step. The resulting vector density was 0.026 valid N_{vpp}. The correct matching rate gives the ratio of correct vectors to detected vectors. Note that all algorithms, with the exception of 3FIT, achieved correct matching rates above 95%.

Table 5.6. Comparison of the number of calculated vectors and the number of valid vectors using polar-coordinate similarity, three-frame minimum acceleration (3FIT), and relaxation (NRX) methods using VSJ PIV standard image #301 (0.06 N_{ppp}) [28].

	Frames employed	Number of vectors (n_{vi})	Number of correct vectors (n_{vci})	Correct matching rate (%)
Original data	0–1	3917	3917	100.0
	0–2	3785	3785	100.0
	0–1–2	7816	7816	100.0
3FIT	0–1	954	896	93.9
	0–1–2	3038	2637	86.8
MF-EPS	0–1	1160	1146	98.8
	0–1–2	3448	3381	98.1
PCSS	0–1	1122	1104	98.4
	0–2	1019	973	95.5
NRX (Ohmi and Li 2000)	0–1	808	788	97.5
	0–2	714	680	95.2
2F-EPS	0–1	1123	1112	99.0
	0–2	1014	977	96.4

5.7 Optimization methods

In this section, a number of tracking algorithms that employ optimization are discussed, which involve minimization of a cost function. The cost function can be as simple as a nearest neighbor criterion or as complicated as a minimization of velocity gradients within the flow. The techniques discussed include the deterministic annealing approach, the variational approach, genetic algorithms, ant colony optimization, and optimization using fuzzy logic.

5.7.1 Deterministic annealing

The deterministic annealing approach is a method for matching particles that attempts to deal with rotational and shearing flows [29]. Essentially, a transformation of particle coordinates from the first frame to the second is determined based on the minimization of a cost function (figure 5.17). The transformation contains both a displacement shift, t, and a mapping matrix, A, that can account for shear and rotation within an interrogation region, which in this implementation was a circular area with a 32-pixel diameter,

$$T_\gamma(X) = Y = t + AX. \tag{5.39}$$

The above transformation is used in the energy cost function

$$\begin{aligned} E(\boldsymbol{m}, y) = \sum\nolimits_{k=1}^{K}\sum\nolimits_{j=1}^{J} m_{kj}\,|Y_k - T_\gamma(X_j)|^2 + g(\gamma) \\ + \alpha\sum\nolimits_{k=1}^{K} m_{k(J+1)} + \alpha\sum\nolimits_{j=1}^{J} m_{j(K+1)}, \end{aligned} \tag{5.40}$$

where m_{kj} is a permutation matrix, which equals one when particle k is the selected match for particle j, α is a constant chosen to determine the maximum tolerable cost of a particle track, and $g(\gamma)$ is a regularization term that can be omitted. The third and fourth terms on the right-hand side of the cost function exist to account for the possibility of no-match particles. Within m_{kj}, a particle in the first frame without a match will be assigned $K + 1$ and a particle in the second frame will be assigned $J + 1$.

Equation (5.40) is minimized iteratively to determine the appropriate transformation for each particle in the first frame. The pseudocode for this minimization

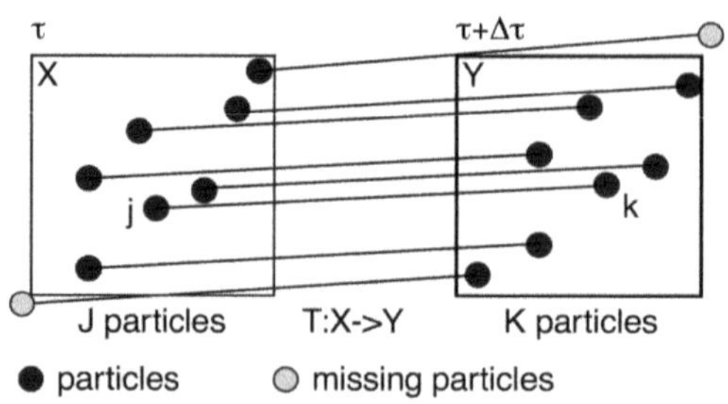

Figure 5.17. Schematic of mapping between frames X and Y. Reprinted from [29] with permission of Springer.

is shown in figure 5.18. There are two steps that are iterated. The first step calculates the probability of each match in matrix m_{kj} and normalizes them. Note that for the minimization, match matrix is not binary, but rather consists of probabilities that are normalized such that the sum equals one in each row and column. The second step recalculates the translation and mapping matrix. There is a parameter, β that is increased exponentially with each iteration, and allows for the convergence of match probabilities in m_{kj} to one. The parameter selection method is thoroughly described by Stellmacher and Obermeyer [29] (figure 5.19).

Deterministic annealing was applied to experimental data at 0.0036 N_{ppp}. The results show that the number of vectors obtained by this approach is comparable to that of super resolution PIV. The resulting particle pairings, however, contain fewer mismatches than nearest neighbor PTV, super resolution PIV, or the relaxation method.

The percentage of incorrectly identified tracks (unsuccessful interrogations) is compared in figure 5.20 for various algorithms using synthetic data. The particle density in this artificial data was 0.0047 N_{ppp}. An unsuccessful interrogation is quantified as any track that results in less than 5% error. For shearing flows, the line for deterministic annealing is not visible, as it lies on the x-axis. The deterministic

BEGIN Outer loop
BEGIN Inner loop
E-Step:
Calculation of the probabilities $\langle m_{kj}\rangle$:
IF $j \neq J+1$ and $k \neq K+1$:
$\langle m_{kj}\rangle = \exp\left(-\beta\,|\mathbf{Y}_k - T_\gamma(\mathbf{X}_j)|^2\right)$
ELSE
$\langle m_{kj}\rangle = \exp(-\beta\alpha)$
Normalization of the probabilities $\langle m_{kj}\rangle$:
BEGIN normalization loop

$$\langle m_{kj}\rangle = \frac{\langle m_{kj}\rangle}{\sum_{j'=1}^{J+1}\langle m_{kj'}\rangle}$$

$$\langle m_{kj}\rangle = \frac{\langle m_{kj}\rangle}{\sum_{k'=1}^{K+1}\langle m_{k'j}\rangle}$$

END Normalization loop
M-Step:
Calculation of transformation parameters:
$\frac{\partial E}{\partial \gamma} \overset{!}{=} 0$
END Inner loop
Increase β
END Outer loop

Figure 5.18. Pseudocode for deterministic annealing minimization. Reprinted from [29] with permission of Springer.

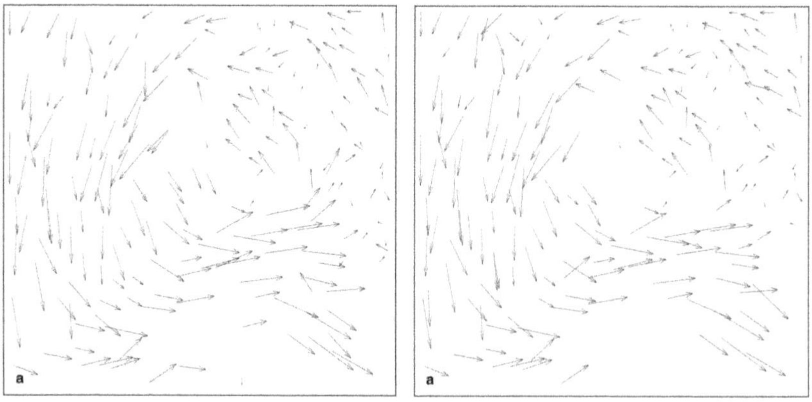

Figure 5.19. Comparison of the deterministic annealing approach (left) and super resolution PIV (right). Reprinted from [29] with permission of Springer.

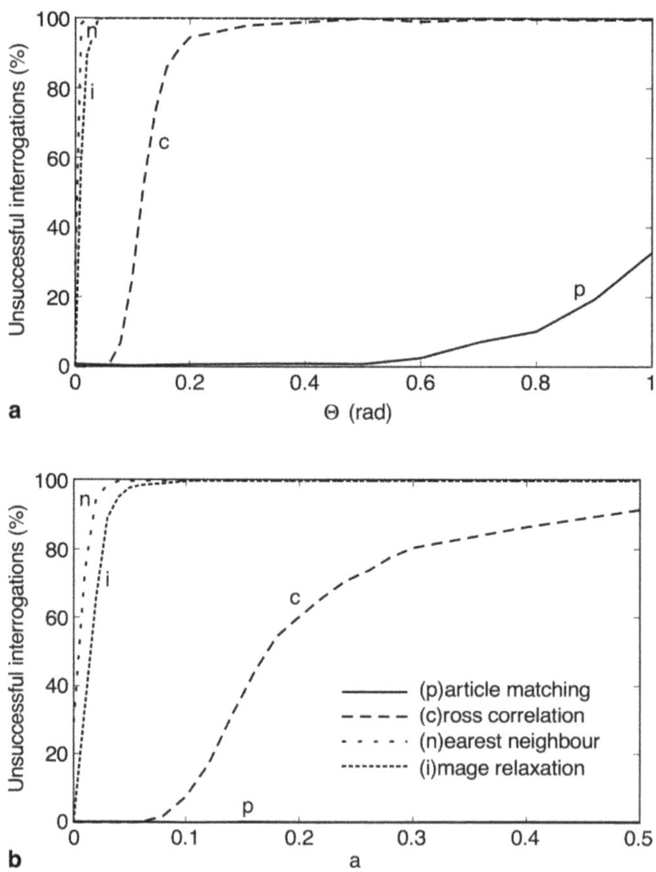

Figure 5.20. Performance of deterministic annealing tracking (here labeled particle matching) for (a) rotational and (b) shearing flows. Reprinted from [29] with permission of Springer.

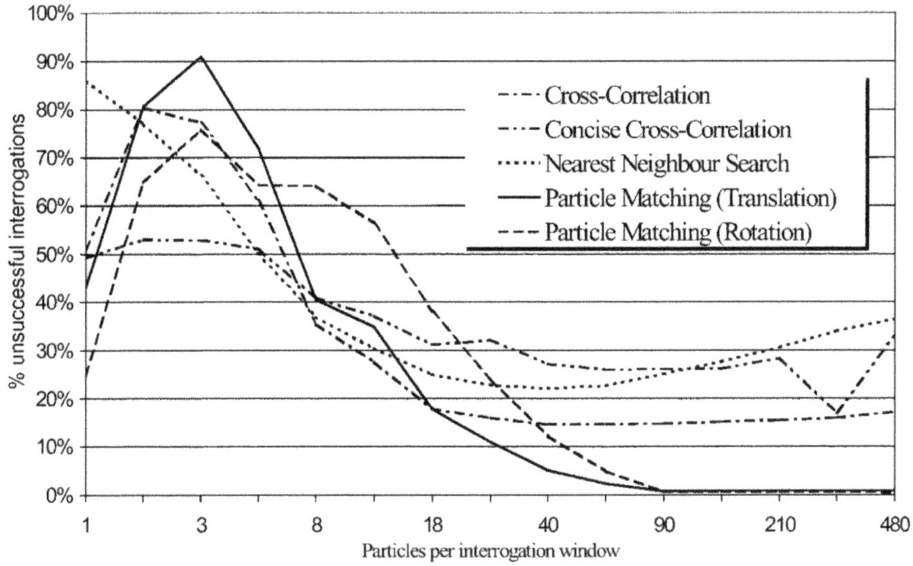

Figure 5.21. Comparison of two deterministic annealing techniques with cross-correlation and nearest neighbor matching techniques in 3D. Reprinted from [30] with permission of Springer.

annealing approach yielded 0.0047 N_{vpp} for the shearing flow test and 0.003 N_{vpp} for flow that rotated one radian in a single time step. The other algorithms failed to recognize more than 0.0002 N_{vpp} for rotations greater than 0.2 radians. These results show the performance benefit of deterministic annealing over cross-correlation, relaxation, and nearest neighbor approaches in the presence of severe velocity gradients.

This algorithm was extended to 3D and applied to holographic PTV by Krepki *et al* [30]. Extension into 3D involves a more complicated 3 × 3 transformation matrix, but otherwise behaves the same as 2D deterministic annealing. In order to test the deterministic annealing tracking technique in 3D, a test was performed using a spherical interrogation volume with diameter 128 pixels. The test had a rotation of 4° and a translation of 12.8 pixels. The seeding density tested for 480 particles was 0.037 N_{ppp}. The algorithms compared are PIV cross correlation, BICC (here labeled concise cross correlation), nearest neighbor, 3D deterministic annealing (labeled particle matching rotation), and simplified 3D deterministic annealing (labeled particle matching translation). Figure 5.21 shows that deterministic annealing performs poorly with a low particle density, but better than the compared algorithms above 0.003 N_{ppp}, corresponding to 40 particles. The nearest neighbor and cross-correlation algorithms showed deteriorating performance at the highest densities tested.

The algorithm was also successfully applied to both laminar and turbulent flows for particle densities ranging from 20 to 350 particles per 3968 × 4032 × 3600 μm^3 interrogation region, although this was not presented in terms of particles per pixels. The computation time for the algorithms was compared for a test with 40 particles using a 333 MHz processor in table 5.7. It can be seen that the deterministic

Table 5.7. Computational speed of tracking algorithms for 40 particle tracks [30].

Interrogation procedure	CPU time (s)		
	Interrogation	Particle extraction	Complete procedure
Cross correlation (64 × 64 × 64)	9.4267	0.00	9.4267
Cross correlation (128 × 128 × 128)	96.523	0.00	96.523
Nearest neighbor	0.0049	0.32	0.3249
CCC	1.2366	0.32	1.5566
PM (translation only)	0.6772	0.32	0.9972
PM (rotation and translation)	3.3724	0.32	3.6924

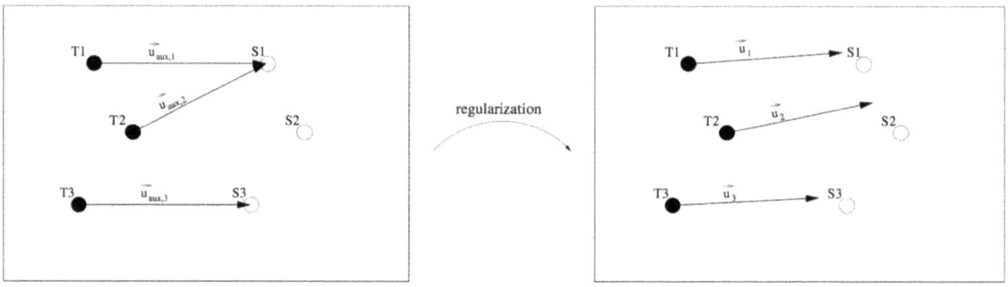

Figure 5.22. Nearest neighbor estimate of particle matches (left) and the result of regularization (right) [31].

annealing technique took twice as long as the BICC method and ten times as long as the nearest neighbor algorithm. The translational-only particle matching algorithm decreased the computation time to less than that of the BICC method. It should also be noted that all PTV techniques were at least three times faster than the PIV cross-correlation techniques.

5.7.2 Variational approach

The variational approach is an iterative method that takes the nearest neighbor concept of particle matching and imposes a smoothness constraint. The technique is described for PTV as developed by Ruhnau *et al* [31]. The mechanism of the variational approach is shown in relation to nearest neighbor estimates in figure 5.22. The smoothness constraint is imposed as a regularization of local velocity vectors.

In order to implement this algorithm, Ruhnau *et al* introduced a convex attraction potential, $E_{\text{local}}(\vec{u})$, which attracts every particle to its closest neighbor

$$E_{\text{local}}(\vec{u}) = \sum_{i=1}^{M} \frac{\alpha}{2}(d_T(S_i + \vec{u}_i))^2, \tag{5.41}$$

where S_i and T_i are the extracted particle coordinates in the first and second images, respectively, $\vec{u}_i$ is the particle displacement, α is a constant that was set to 0.8 in this test, and the operator d_T gives the Euclidean distance between its argument and the nearest particle position T in the second image for particle i, and S_i is spatial coordinate in the first frame. The operator $d_T(S_i,T)$ is the Euclidean distance between a specific particle S_i and the nearest particle within the particle set T in the second image. This minimization alone gives nearest neighbor matching for all the M particles in the set S.

The variational approach also applies a smoothness constraint, which is given as a minimization of velocity gradients within the flow. This term is added with a scalar multiplier, λ, acting as a smoothness parameter, to create the variational framework. The selection of the scalar multiplier allows the user to determine how much weight the smoothness constraint should hold in the minimization, and in this paper it was set to 0.1. The smoothness constraint is

$$E_{\mathrm{global}}(\vec{u}) = \int_{\Omega} \sum_{j=1}^{N} |\nabla \vec{u}^{j}|^{2}\, ds, \tag{5.42}$$

where N is the number of dimensions, and Ω is the spatial domain. The two constraints are combined as

$$E(\vec{u}) = E_{\mathrm{local}}(\vec{u}) + \lambda E_{\mathrm{global}}(\vec{u}), \tag{5.43}$$

where minimization is performed iteratively by deforming the current vector field, then regularizing it with respect to neighbors. Matchless particles are removed if their contribution to the minimization exceeds an adjustable threshold. Iterations are done to minimize the combination of potentials until the solution converges.

A comparison of the variational approach with others is given in table 5.8. The test was done on VSJ standard image 301 using the particle mask correlation method for identification, resulting in roughly 0.017 N_{ppp}. The table includes the number of found vectors, the number of valid vectors, and the reliability (ratio of estimated and correct matches) of a few algorithms. The variational approach achieved 0.014 valid N_{vpp} in the test. It yielded both the most vectors and highest reliability compared to four-frame, binary cross-correlation, and relaxation algorithms.

In a 3D extension of this method (shown in table 5.9), tests of the algorithm using VSJ standard image 331 performed with yield and reliability both exceeding 98% except for a severe decline when matching the first and third frames with the percent

Table 5.8. Comparison of four-frame tracking (FIT), binary image cross correlation (BCC), relaxation (NRX), and variational approach (VAR) for the VSJ 301 image sequence [31].

Algorithm	Frames	Estimated matches	Correct matches	Reliability(%)
FIT	0, 1, 2, 3	630	559	88.73
BCC	$0 \rightarrow 1$	860	788	91.62
	$0 \rightarrow 2$	863	691	80.07
NRX	$0 \rightarrow 1$	808	788	97.52
	$0 \rightarrow 2$	714	680	95.24
VAR	$0 \rightarrow 1$	872	865	99.2
	$0 \rightarrow 2$	904	885	97.9

Table 5.9. Testing of the variational approach for VSJ standard image 331 [31].

	00 → 01			00 → 02		
Removed panicles (%)	Possible matches	Yield (%)	Reliability (%)	Possible matches	Yield (%)	Reliability (%)
0	3364	100.00	99.76	3192	99.97	99.47
5	3037	100.00	99.84	2881	99.86	99.45
1	2731	100.00	99.60	2586	99.38	99.34
15	2440	100.00	99.59	2307	98.22	99.60
20	2170	100.00	99.40	2053	98.30	99.56
25	1885	100.00	99.74	1809	44.83	44.81
30	1649	100.00	99.40	1557	38.79	39.35
35	1403	99.93	99.64	1339	31.14	31.17
40	1211	100.00	99.26	1131	32.98	33.01

of removed particles exceeding 25%. The seeding density of this 3D synthetic data was roughly 0.05 N_{ppp}, and when particle images were not removed, the variational approach yielded roughly 0.05 valid N_{vpp}. Note that the yield listed defines the ratio of identified matches to possible matches.

5.7.3 Genetic algorithms

A genetic algorithm (GA) works to achieve a goal while minimizing a constraint, or fitness function. Genetic algorithms have been applied to PTV by Oyama and Kaneko [32], Sheng and Meng [33], and Doh *et al* [34] by creating a fitness function that minimizes the sum of particle displacements between two frames. This algorithm works for low particle densities, but performance deteriorates when the mean particle spacing approaches the particle displacement ($\xi \rightarrow 1$). Sheng and Meng [34] tracked 215 particles (0.000 87 N_{vpp}) in an experimental holographic PTV experiment.

Kimura *et al* [35] and Furukawa *et al* [36] based the fitness function on the morphology of neighboring particles, which successfully increased the possible number of tracked particles by three to four times, but never more than 1000 particles (0.015 N_{vpp}). Ohmi and Yoshida [37] based their fitness function on the rigidity of particle cluster patterns, which achieved more than 0.03 N_{vpp}.

The following is the genetic algorithm proposed by Ohmi and Panday [38], which used the same fitness function as Ohmi and Yoshida [37] with changes made only to the genetic mutations in order to increase convergence. The first step for creating a genetic algorithm is defining a genetic encoding. In this case, the genetic encoding is in the form of chromosomes. Each chromosome is a list of particle IDs for a single frame. An individual is a set of two chromosomes, one for each image. The corresponding elements of each chromosome indicate particle pairs (i.e. the particle listed in the *i*th element of the first chromosome is paired with the particle listed in

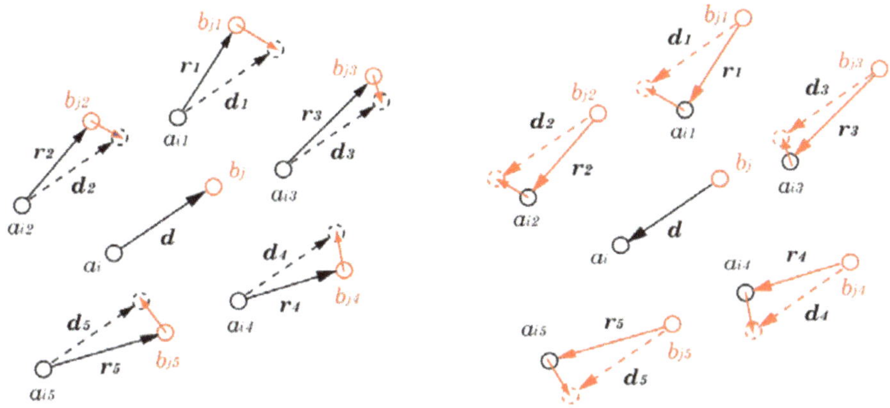

Figure 5.23. Relaxation parameter used as a fitness function by Ohmi and Panday. Reprinted from [38] with permission of Springer.

the *i*th element of the second chromosome). The two chromosomes do not need to be the same length, allowing for no-match particles to exist. The individuals mutate and reproduce with one another using a number of methods, which allows for production of new individuals. The fitness is evaluated to determine which individuals continue to reproduce,

$$F = \sum_{i=1}^{m} \sum_{k=1}^{p} |d_k - r_k|, \tag{5.44}$$

where F is the fitness function defined to evaluate the genetic algorithm. There are m particles in the first image, with p neighbors for each. d_k is the virtual displacement of particle k, which is equal to the displacement of the particle m. r_k is the actual displacement of particle k (figure 5.23).

The genetic operations performed in this technique include the generation of the initial population, selection and reproduction, sorting, crossover, and mutation. The generation of the initial population is simply a randomized integer labeling of the particle IDs in both image one and image two. Selection is the step of the GA in which the fitness function is evaluated for each individual. Selection is done using either exponentially ranking selection, which generates child individuals at a rate determined by the exponential ranking of their fitness evaluation, or unconditional elite-reproducing selection, which reproduces from the most-fit individual in the generation history at a prescribed rate. When used with gene code sorting, both methods work effectively if used with the proper crossover and mutation methods. Sorting of genes is done every 10 to 50 generations to ensure that the particles that have good fitness scores are less likely to change than those with poor scores. The individual gene code pairs have their portion of the fitness function calculated, and the lowest scores are moved to the front of the chromosomes. Crossover and mutation are suppressed in the front part of the chromosome to protect the good gene code pairs from being separated. The crossover step uses two parents and splices their chromosomes at certain points to create offspring. The splicing of genes

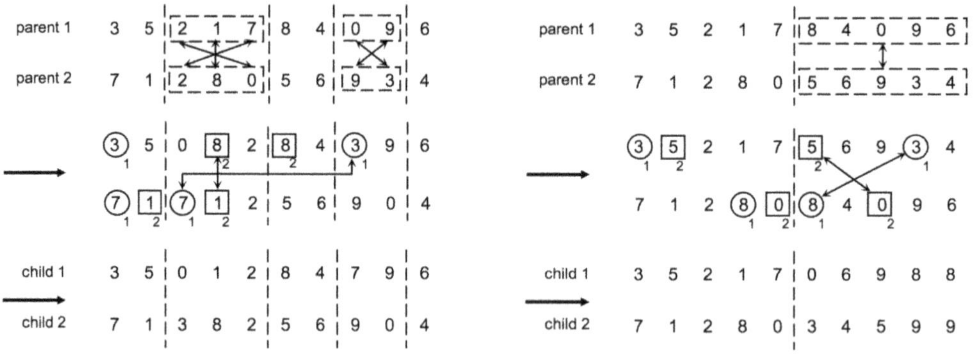

Figure 5.24. Multi-point inverse crossover (left) and single-point partially matched crossover (right). Reprinted from [38] with permission of Springer.

can cause two of the same particle to occur in the child chromosomes, and as such duplicates are moved accordingly. Figure 5.24 schematically shows this crossover process.

Mutation occurs randomly, selecting two genes in a single chromosome to switch. The mutation rate is fixed throughout the algorithm. The above genetic operations are all applied iteratively until an individual reaches a converged fitness value and the gene code pairs stabilize. These gene pairs are then taken as the final particle match results.

Using the PIV standard image #301 for a 2D test, GA_PTV calculated 0.031 N_{vpp} (from a 0.031 N_{ppp} image) with no spurious vectors. The results from this test are shown in figure 5.25. It was found that for 2D data, this particle concentration was the upper limit for achieving a correct tracking rate of 100%. The highest tested possible N_{vpp} was 0.034, and the algorithm achieved 99.7%, so the seeding density could be further increased. The algorithm also successfully matched 3D PTV results for 2000 particles (0.0076 N_{ppp}) with 100% accuracy using exact particle locations in 3D PIV standard image #351. Using a 1.8 GHz Core2Duo PC the computation took 474 s to achieve the 2000 matches.

5.7.4 Ant colony optimization

Another algorithm for particle tracking is based on ant colony optimization. This is the same technique that is described for stereoscopic particle location in section 4.2.1.3 aside from the fitness function used, which was used to update the pheromone for a particular particle track. Its use for particle tracking was introduced by Takagi [39] and improved by Ohmi *et al* [40]. In the context of tracking particle motions, agents investigate particle correspondences at two time steps. As a nearest neighbor search is not a reliable metric for high seeding densities, the fitness function in this case is based upon the relaxation distance. The pheromone update is

$$\tau(i, j) \leftarrow (1 - \rho)\tau(i, j) + \sum_{k=1}^{m} \Delta\tau^{k}(i, j), \tag{5.45}$$

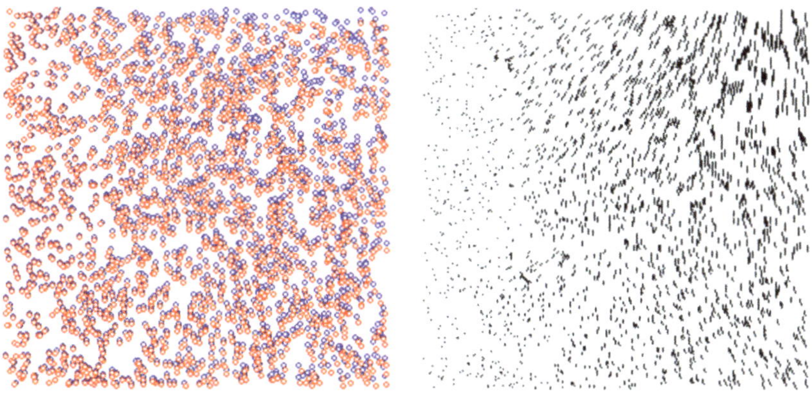

Figure 5.25. Original particle locations (left) and particle matching result (right) using the genetic algorithm for PIV standard image #301. Reprinted from [38] with permission of Springer.

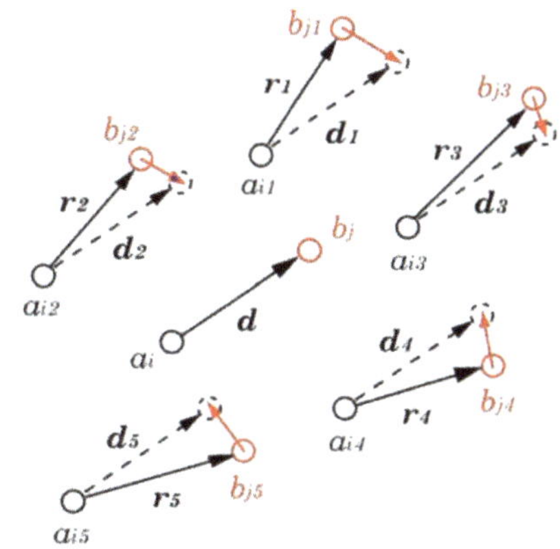

Figure 5.26. Concept of relaxation lengths for ant colony optimization. Reprinted from [40] with permission of Springer.

and

$$\Delta\tau^k(i, j) = \begin{Bmatrix} 1/L^k & \text{if } (i, j) \in T^k \\ 0 & \text{otherwise} \end{Bmatrix}, \tag{5.46}$$

where ρ is a constant that determines the dissipation rate of pheromone and L^k is the new relaxation length given by

$$L^k = \sum\nolimits_{i=1}^{m} \sum\nolimits_{k=1}^{p} |d_k - r_k|, \tag{5.47}$$

where d_k is the displacement from particle i to j and r_k is the displacement estimate of nearby particles. This is essentially the same as the relaxation probability from section 5.3 and is shown in figure 5.26, where b_{jk} is the length that is summed and minimized.

This algorithm is referred to as the relaxation minimization algorithm for comparison of results. The comparison of this algorithm with a four-frame tracking approach and the binary cross-correlation approach is given below in table 5.10.

Four test cases were done for each technique, two with 500 potential pairs, two with 1000 potential pairs. The ant colony optimization algorithm outperformed both the four-frame and BCC tracking methods in all the tests, achieving roughly 0.0152 N_{vpp} (99.8% correctly identified tracks) in the 1000 particle pair case (0.0153 N_{ppp}). The algorithm was also tested for 3D synthetic data using VSJ image #351 (0.008–0.015 N_{ppp}) and achieved 100% yield and reliability. This increase in performance is due to the fact that tracking is easier in 3D than in 2D if accurate 3D locations have been determined.

5.7.5 Fuzzy logic PTV

Fuzzy logic algorithms employ a tolerance for imprecision while performing an optimization. This concept is often used in control systems; however, identifying particle tracks with noisy data is a good candidate for fuzzy control. This was applied to PTV by Wernet [41, 42]. In order to determine particle tracks using fuzzy logic, two frames must be acquired, and particle centroids must be located. Centroids in frame 1 are used as starting points and all of the centroids in frame 2 within a maximum displacement are initially considered. At this point, the vector field is noisy and there are many contradictory particle tracks, as shown in figure 5.27.

For particles in frame 1 that have common matches in frame 2, then all possible matches are compared. The assumption is that if two particles are close enough to share a potential match, then their displacements should be similar. This holds true for higher order interactions as well. There are four inputs for each vector pair: the distance between vector midpoints in pixels (Sep), average vector magnitude (Mag), difference in vector magnitude (MagDif), and the sum of squares of the differences

Table 5.10. Comparison of ant colony optimization using the relaxation minimization to other PTV tracking algorithms using VSJ image #301 (0.008–0.015 N_{ppp}) [40].

Frame	Name of the PTV methods	Number of particles	Correct pairs	Correct rate (%)
0–1–2–3	Four-frame in-line tracking	500	459	91.8
		1000	814	81.4
0–1	Binary cross correlation	500	382	76.4
		1000	869	86.9
	Relaxation minimization	500	100	100
		1000	998	99.8
0–1–2–3	Four-frame in-line tracking	500	459	91.8
		1000	814	81.4
0–2	Binary cross correlation	500	283	56.6
		1000	659	65.9
	Relaxation minimization	500	498	99.6
		1000	985	98.5

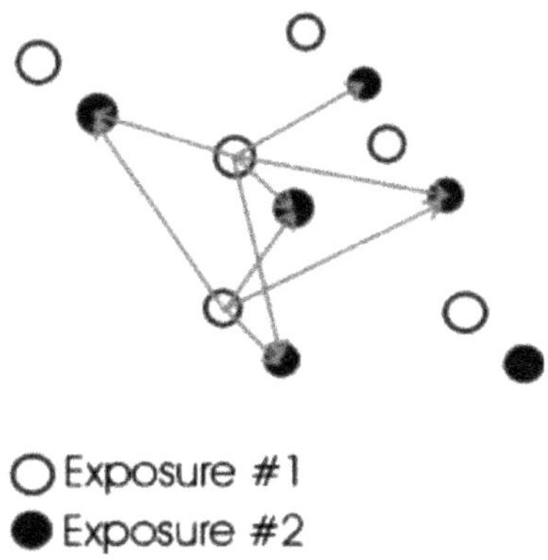

Figure 5.27. Particle centroids from two exposures superimposed. Potential candidates are shown for each particle in exposure 1 with a vector. Reprinted from [42] with permission of Springer.

of the x and y components of the two velocity vectors (Delta). All of these inputs are processed based on the confidence outputs shown in table 5.11. Input combinations are all classified as the combination of high, medium, and low. The combination is quantified as a membership to each category, μ. The concept of membership is shown in figure 5.28. The range for each is determined by a minimum and maximum possible value (for example, the minimum and maximum displacements are 0 and 20 pixels, respectively, in figure 5.28). The confidence outputs have values of 0.1, 0.5, and 0.8 for low, medium, and high, respectively. The confidence outputs from each category are weighted by a vector pair's membership and summed as

$$\text{Confidence} = \frac{\sum_{i=1}^{N} \text{Conf\_out}_i^* \mu\_\text{out}_i}{\sum_{i=1}^{N} \mu\_\text{out}_i}, \tag{5.48}$$

where N is the number of outputs.

After a single pass of this confidence output, the fuzzy logic algorithm is applied again to compare non-interacting velocity vectors. The resulting vector with the maximum confidence is selected for each particle in frame 1. The final confidences that exceed a threshold value are then considered valid vectors.

Wernet [42] suggested using PIV results as a guide for tracking. Cross-correlation results give an initial estimate of local velocity within each interrogation region. For each first frame particle, the nearest four cross-correlation vectors are used for comparison using the fuzzy logic described above, resulting in a particle match that is most similar to the local cross-correlation results. This is repeated for each first frame particle.

In order to test this tracking algorithm, experimental data imaging a centrifugal compressor with impeller tip speed equal to 490 m s^{-1} was used. Within the 1000×1000 pixel CCD, 6400 particle images (0.0064 N_{ppp}) were detected in each frame. Using the PIV results to guide fuzzy logic tracking, 4000 velocity vectors

Table 5.11. Confidence outputs based on interaction of fuzzy logic inputs [42].

	MagDif Small			MagDif Med			MagDif Large		
	Mag Small	Mag Med	Mag Large	Mag Small	Mag Med	Mag Large	Mag Small	Mag Med	Mag Large
					Sep Small				
Delta Small	High	High	High	High	Med	Med	Med	Med	Low
Delta Med	High	Med	Low	Med	Med	Med	Low	Low	Low
Delta Large	Med	Low	Low	Low	Low	Low	Low	Low	Low
					Sep Med				
Delta Small	High	Med	Med	Med	Med	Low	Med	Low	Low
Delta Med	Med	Med	Low	Med	Med	Low	Low	Low	Low
Delta Large	Low	Low	Low	Low	Low	Low	Low	Low	Low
					Sep Large				
Delta Small	Mtd	Med	Low	Med	Low	Low	Low	Low	Low
Delta Med	Low	Low	Low	Low	Low	Low	Low	Low	Low
Delta Large	Low	Low	Low	Low	Low	Low	Low	Low	None

(0.004 N_{vpp}) were tracked, which is equal to a yield of over 60%. Of the 4000 vectors, 294 were identified as spurious because they deviated from the mean by 45% while the turbulent fluctuations were expected to be no more than 15%, resulting in a reliability of 93%.

5.8 Delaunay tessellation methods

Methods in this category match patterns using Delaunay tessellation in order to find particle matches in PTV (DT-PTV). The DT-PTV techniques are based on an algorithm by Song *et al* [43]. Delaunay tessellation takes a set of scattered points in space and generates a set of triangles. These triangles are generated such that there are no nodes (scattered points) that exist in the circumcircle of any tessellated triangle. The technique for generating a Delaunay tessellation begins by generating a super-triangle (shown in figure 5.29(a)) that encompasses all of the nodes. Within the super-triangle, nodes are introduced one at a time. Whichever triangle a new node

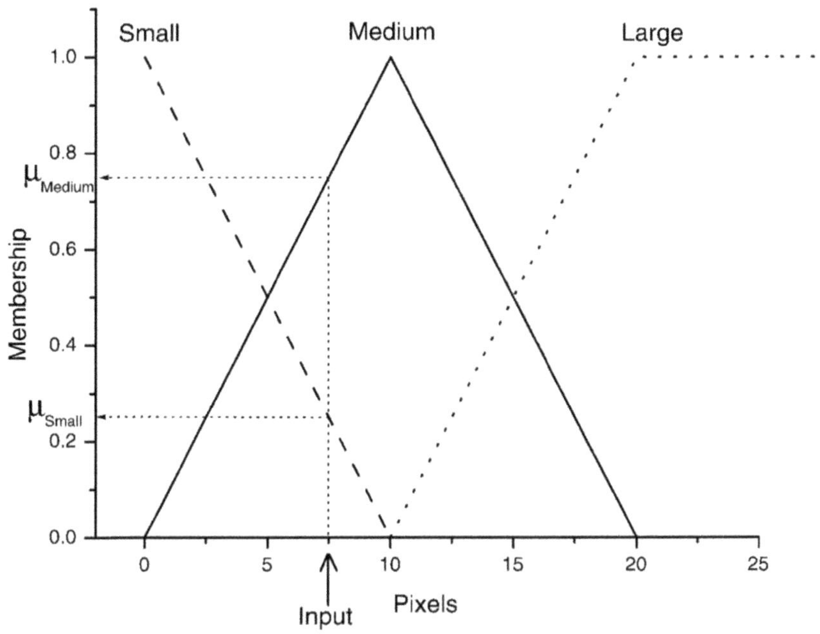

Figure 5.28. Representation of membership based on a fuzzy logic input. Reprinted from [42] with permission of Springer.

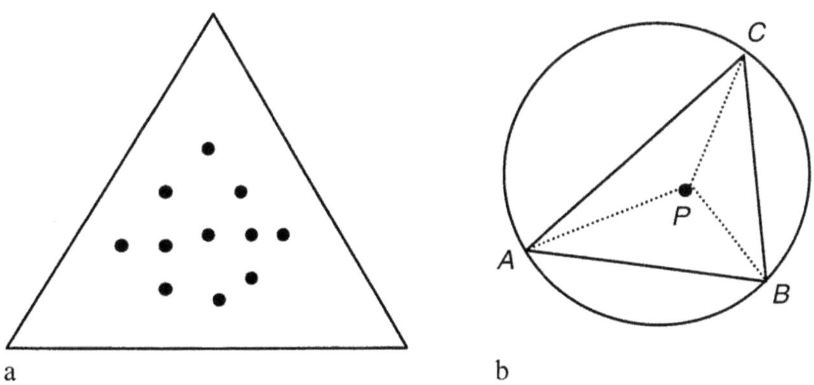

Figure 5.29. (a) Super-triangle and (b) triangle division for a new point, *P*, introduced in Delaunay tessellation. Reprinted from [43] with permission of Springer.

intersects with is split into three different triangles. This super-triangle and subsequent division is shown in figure 5.29(b). This is repeated until all nodes have been introduced. In most cases, there exists one unique tessellation for a given set of scattered nodes.

In order to apply Delaunay tessellation to particle tracking, a tessellation is generated for each of two consecutive images, where each particle location is a node. After tessellation, the triangles are each indexed. For each triangle in the first frame, a cross-correlation coefficient is calculated to find the overlapping area with each triangle in the second frame within a maximum displacement distance. This coefficient is calculated as

$$C_{ij} = \frac{\text{Area}(\text{tri}_i \cap \text{tri}_j)}{\sqrt{\text{Area}(\text{tri}_i)\text{Area}(\text{tri}_j)}}, \tag{5.49}$$

where the overlapping area of triangles can be calculated using the method named 'clipping of polygons', described by Harrington [44]. The overlap area is defined based on the number of vertices used. This can range from 3 to 6 for two overlapping triangles. Thus, the area of a polygon with an even number of vertices Np is

$$S = \frac{1}{2}\sum_{k=2}^{Np/2}\{(x_{2k-1} - x_1)(y_{2k} - y_{2k-2}) + (x_{2k} - x_2)(y_{2k+1} - y_{2k-1})\}, \tag{5.50}$$

and the area for a polygon with an odd number of vortices is

$$S = \frac{1}{2}\sum_{k=2}^{Np}(x_{\text{k}} - x_1)(y_{\text{k}+1} - y_{k-1}), \tag{5.51}$$

where x_k and y_k denote the coordinates of vertex P_k.

In order to account for flow deformations, translation and rotation of triangle areas must be applied. Song's algorithm translates the center of gravity of the two triangles to be compared to the same point, then both triangles are rotated such that the vertex of the smallest angle of each triangle lies on the positive x-axis. This translation and rotation procedure is shown for two triangles in figure 5.30.

The match for a triangle in image 1 is selected as the triangle in image 2 corresponding to the maximum correlation coefficient. The individual particle matches are then selected as the nearest vertices in the rotated and overlapped triangles.

This technique was tested and compared to BICC for simulated analytical Taylor–Green vortices with a Reynolds number of 1000. The results are shown in table 5.12 for a range of particle numbers, from 0.01 N_{ppp} to 0.023 N_{ppp}. For low density images with small deformations the DT-PTV algorithm performed no better than BICC. However, in the tests using larger time steps between frames, and therefore containing more substantial flow deformations, DT-PTV managed to increase the vector yield by 42% while also reducing the computation time by roughly two orders of magnitude compared to BICC.

Zhang *et al* [45] made two improvements to DT-PTV that allowed the approach to perform in three dimensions. The rotation and volume overlap calculation of tetrahedrons involves tedious, although not impossible, algorithms by introducing a characteristic vector to describe each tetrahedron. The characteristic vector for a tetrahedral includes a value for the area of each face of the tetrahedron. These face areas are sorted by increasing area. The tetrahedrons in the first and second images must be sorted the same order, so as to simulate rotation. The correlation coefficient between two tetrahedrons is

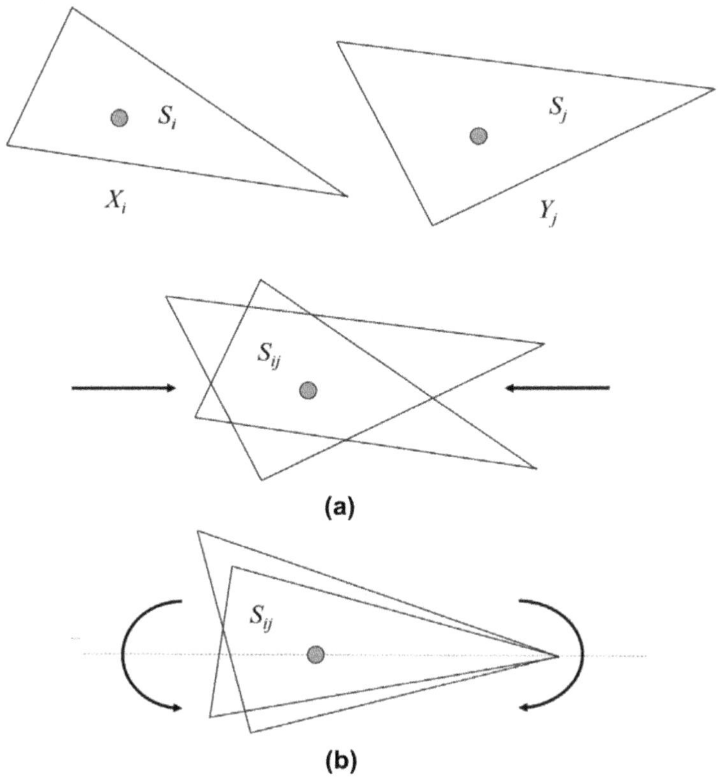

Figure 5.30. Translation and rotation of polygons for coefficient calculation [43].

Table 5.12. Comparison of computation time and number of vectors obtained for the BICC and DT-PTV tracking algorithms [43].

Num.	Δt	R	CPU time		Vectors num.	
			BICC	DT-PTV	BICC	DT-PTV
632	0.01	0.1	131	6	632	632
631	0.05	0.1	130	6	630	631
1505	0.1	0.12	2480	24	1500	1480
1498	0.2	0.22	7437	26	1245	1383
1500	0.3	0.33	14 721	31	910	1295

$$C_{ij} = \left[\frac{\min\left(|X_i'|, |Y_j'|\right)}{\max\left(|X_i'|, |Y_j'|\right)}\right] * \left[\frac{\mathrm{cov}\left(X_i', Y_j'\right)}{\sqrt{\mathrm{var}(X_i')\mathrm{var}\left(Y_j'\right)}}\right], \quad (5.52)$$

where X_i' is the sorted characteristic vector of the tetrahedron in the first frame and Y_j' is the characteristic vector of the candidate tetrahedron in the second frame.

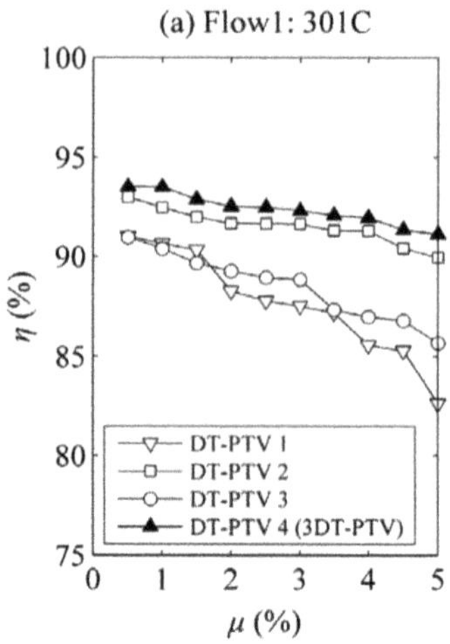

Figure 5.31. Comparison of DT-PTV methods with the improvements suggested by Zhang *et al* for VSJ standard image #301 (~0.06 N_{ppp}) [45].

The second change was to use tessellation to eliminate the maximum displacement criteria for the search area. Instead, triangles in the second frame are organized such that for triangle D_i, all triangles sharing a vertex are considered a first connection. Then all triangles sharing a vertex with these are considered a second connection to D_i and so on. A maximum number of connections, rather than a maximum displacement, is then selected such that all triangles in frame 2 that are within α connections of the reference triangle in frame 1 are considered potential matches.

The algorithm also uses a dual computation technique (see section 5.11) that calculates matches going from frame 1 to 2 and from frame 2 to 1. In this way, there are redundant matches found for each particle pair. When two particles claim the same match, the pair with the most occurrences are selected as the correct match. This is the same technique used by Jia *et al* [16] in section 5.3.

DT-PTV was tested with and without the above improvements. DT-PTV 1 used a fixed search radius and the net-flux method, DT-PTV2 used a fixed search radius and the dual computation method, DT-PTV 3 used the DT search with only first connection triangles and the net-flux method, and DT-PTV 4 used the DT search and dual computation. Figure 5.31 compares the performance of the different DT-PTV algorithms using VSJ standard image 301 with 0.06 N_{ppp}. DT-PTV 4 achieved the highest ratio of correct matches to possible matches, η, for the flow investigated as a function of the particle dropout rate, μ. It is clear that the dual computation method substantially improved the results of the DT-PTV algorithm for high particle dropout rates. Given the particle density of 0.06 N_{ppp}, the vector density obtained by DT-PTV 4 for $\eta = 91\%$ and $\mu = 5\%$ was 0.053 N_{vpp}.

5.9 Voronoi diagram methods

Another technique, called VD-PTV, was introduced by Zhang *et al* [46] and is based on the Voronoi diagram (VD). Voronoi diagrams are similar to Delaunay tessellations; however, the Voronoi diagram segments an image such that a single cell exists that surrounds each data point. For a given data point, its corresponding cell is a region that comprises all points closer to that data point than others. In this case, a data point is a particle image. The Voronoi diagram is generated by creating a Delaunay tessellation, generating a circumcircle for each triangle, identifying the midpoint of each circumcircle, and connecting them in a nearest neighbor sense. An example VD is shown in figure 5.32.

For each particle image, the polar radius of the Voronoi cell is found as a function of polar angle. This is called the characteristic curve and is shown in figure 5.33. The curve for a Voronoi cell is defined as the polar radius, r, as a function of polar angle, α. The curve is determined by first dividing the cell into triangles and calculating

$$r(\alpha) = \frac{h}{|\sin(\alpha + \theta_1 - \alpha_1) + 1 - \text{logical}(\theta_1 - 90°)|}, \tag{5.53}$$

where

$$\text{logical}(x) = \begin{matrix} 1 & (x \neq 0) \\ 0 & (x = 0) \end{matrix}, \tag{5.54}$$

and

$$h = \frac{|(y_2 - y_1)x_0 + (x_1 - x_2)y_0 + x_2y_1 - x_1y_2|}{\sqrt{(y_2 - y_1)^2 + (x_1 - x_2)^2}}, \tag{5.55}$$

where (x_i, y_i) are the coordinates of each vertex in a triangle and θ_1 is the angle at vertex 1. The curve for each triangle in a Voronoi cell is then combined piecewise such that the polar radius is defined for α between 0 and 360 degrees.

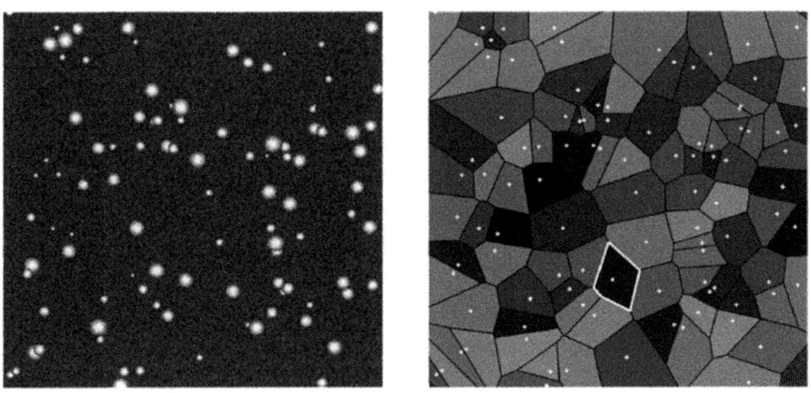

Figure 5.32. Characteristic curve of a Voronoi cell. (left) Original particle image. (right) Voronoi diagram of this image in which the original particles are denoted by white dots [46].

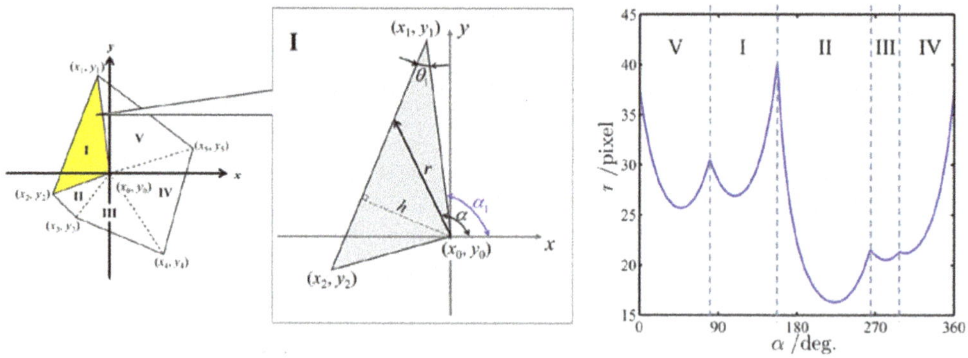

Figure 5.33. The development of the Voronoi cell characteristic curve. (left and middle) Division of a selected cell into a group of triangles (I, II, …,V). In triangle I, the relationship between polar radius r and polar angle α is established according to equations (5.53)–(5.55). (right) The selected cell is converted into its characteristic curve $r(\alpha)$. The curve moves leftwards to fit in the range [0°, 360°] [46].

In the proposed method for particle matching, the characteristic curves for reference particles and candidate particles are compared. A coefficient for the similarity between the curves is calculated and the particle images with the highest similarity are chosen as a match. The similarity coefficient is calculated as

$$C_r = \frac{\mathrm{cov}(r_1, r_2)}{\sqrt{\mathrm{cov}(r_1)\mathrm{cov}(r_2)}}. \tag{5.56}$$

The similarity coefficient is calculated between $r_1(\theta + \alpha)$ and $r_2(\alpha)$, where θ ranges from 0 to 360 degrees, which allows for rotation in the matching process. The maximum C_r calculated is then taken as the match coefficient between the reference and candidate particle. The candidate particle with the highest associated similarity coefficient is selected as the match for every reference particle. If two candidate particles share the highest similarity coefficient, then the reference is considered to have no match. If two reference particles claim the same candidate particle, then the one with the higher similarity coefficient is considered the match.

A comparison of VD-PTV with relaxation and DT-PTV was performed using a synthetic flow consisting of a superposition of shear flow, a vortex, and a dipole. The image was 256 × 256 pixels and the mean particle displacement was 2.0 pixels. The particle concentration was varied from 4 to 25 N_{ppp}. Note that this particle concentration is not possible in experiments, and the most important parameter to consider is the ratio of mean particle displacement to mean particle spacing, $\overline{C_{\mathrm{PTV}}}$. Figure 5.34 shows that VD-PTV has the highest correct matching rate, η, at the shown ranges of particle densities and search radii, R_s. Note that while the relaxation DT-PTV algorithms performed well when the search radius was kept near the mean particle displacement, performance deteriorated when the search radius increased. VD-PTV shows that it is effective even when the mean particle displacement is not explicitly known. Furthermore, the computation time for VD-PTV was less than half of that for DT-PTV and less than 15% of that for relaxation.

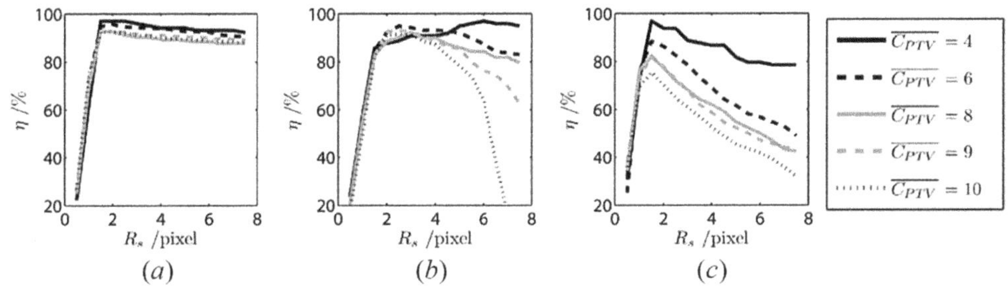

Figure 5.34. Comparison of (a) VD-PTV, (b) relaxation, and (c) DT-PTV with % proper matches, η, versus the search radius, R_s. $\overline{C_{\mathrm{PTV}}}$ is the ratio of average particle displacement to average particle spacing [46].

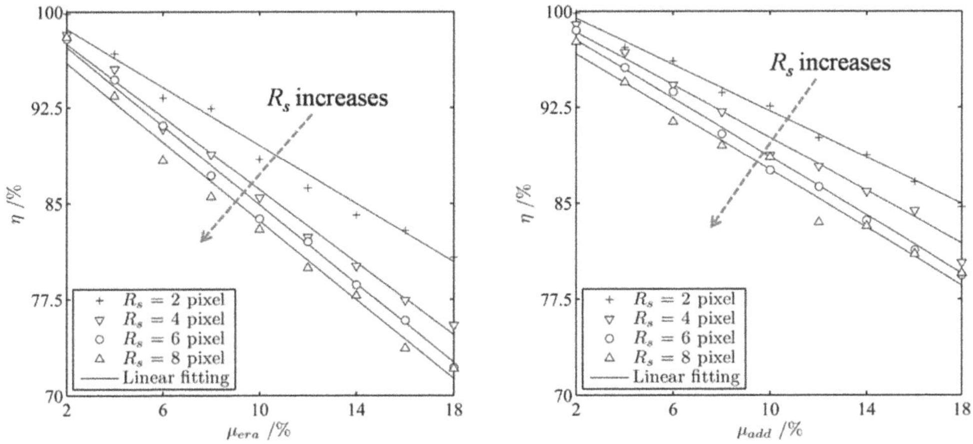

Figure 5.35. Influence of erased (left) and added (right) particles on the performance of VD-PTV for the homemade flow with $\overline{C_{\mathrm{PTV}}} = 8$ [46].

Using the synthetic flow with $\overline{C_{\mathrm{PTV}}} = 8$, the effect of adding and removing particles was tested. The fraction of erased particles, μ_{era}, and added particles, μ_{add}, were each varied from 2% to 18%, the results of which are shown in figure 5.35. The correct matching rate decreased linearly with increasing μ_{era} or μ_{add}. This is due to the fact that adding or removing a particle affects the Voronoi cells of nearby particle images.

5.10 Vision-based PTV

The vision-based matching algorithm, introduced by Lei *et al* [47] in 2012 and improved upon by Paul *et al* [48] in 2014, correlates image features by adhering to three principles: proximity, similarity, and exclusion. The method was first introduced by Scott and Longuet-Higgins in 1991 [49]. In order to enforce the principle of proximity, the proximity matrix, G, is created based on r_{ij}, the distance from point i in frame 1 to point j in frame 2,

$$G_{ij} = e^{-r_{ij}^2/2\sigma^2}, \tag{5.57}$$

where σ is a characteristic distance, originally selected to be the maximum expected displacement. Singular value decomposition is then performed on the new G matrix (equation (5.58)). The two rotation matrices T and U are maintained while the singular value matrix, D, is replaced with the identity matrix of the same dimensions. This normalizes to give the proximity matrix, P, which has a least squares mapping between i and j (equation (5.59)). Within the P matrix, if element i, j is the maximum value in both the ith row and the jth column, then the two particles are considered a match:

$$G = TDU \tag{5.58}$$

$$P = TIU. \tag{5.59}$$

In 1997 Pilu [50] proposed a modified method that resulted in improved particle pair matching, where feature similarities were accounted for using a normalized cross-correlation matrix, C_{ij}, (equation (5.60))

$$C_{ij} = \frac{\sum_{u=1}^{W}\sum_{v=1}^{W}(I_A(u, v) - I_{Am})(I_B(u, v) - I_{Bm})}{W^2\sum I_A \sum I_B}, \tag{5.60}$$

where $I_A(u, v)$ and $I_B(u, v)$ are the pixel intensities in image A and B, respectively, and I_{Am} and I_{Bm} are the mean pixel intensities in each frame. The cross-correlation matrix is then used to create an updated proximity matrix

$$G_{ij}' = e^{-r_{ij}^2/2\sigma^2} \cdot e^{-(C_{ij}-1)^2/2\gamma^2}, \tag{5.61}$$

where γ is a factor that controls the speed of decay of the weighting of the cross-correlation term. It was set to 0.4 for this algorithm.

The match is then tested using an outlier detection scheme (see section 5.11) to validate the match. If valid, the row and column are removed from P in order to maintain the exclusivity principle. If the match is identified as an outlier, the coefficient $p_{i,j}$ is then set to zero and the next maximum is selected. This scheme is iterated until all particle pair matches are achieved.

The algorithm used by Scott and Longuet-Higgins selected σ to be the average displacement of the particles. The improvement made by Pilu suggested that the PIV result for displacement estimate should give the σ value used. Lei *et al* suggested the use of twice the PIV displacement for σ, as tracking performance is improved when σ is an overestimate of the displacement.

Further improvement by Paul *et al* modified the proximity matrix as

$$G_{ij} = \exp-\left[\frac{(r_{ij} - r_{\mathrm{PIV}})^2}{2\varepsilon^2}\right] * \exp\left\{\left[\left(\tan^{-1}\frac{dy_{ij}}{dx_{ij}}\right)^2 - \left(\tan^{-1}\frac{dy_{\mathrm{PIV}}}{dx_{\mathrm{PIV}}}\right)^2\right] \Big/ 2\varphi^2\right\}, \tag{5.62}$$

where it was suggested that ε be set to 4 and φ be set to 1 for high gradient flows. This modification favors particles with a displacement equal to the PIV results over particles with zero displacement. Figure 5.36 shows the modification graphically.

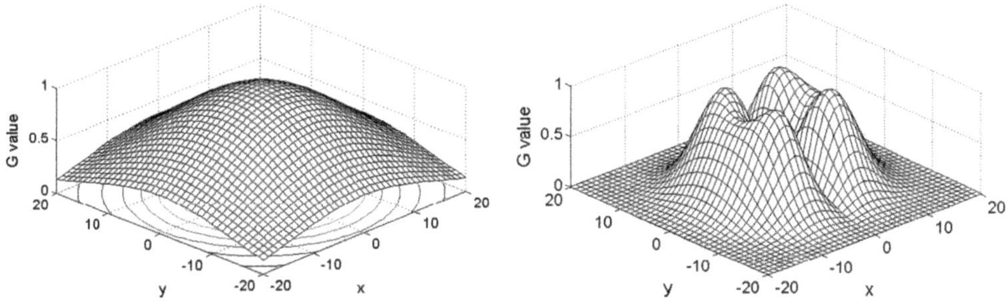

Figure 5.36. Proximity surface with (right) and without (left) modifications made by Paul *et al.* Reprinted from [48] with permission of Springer.

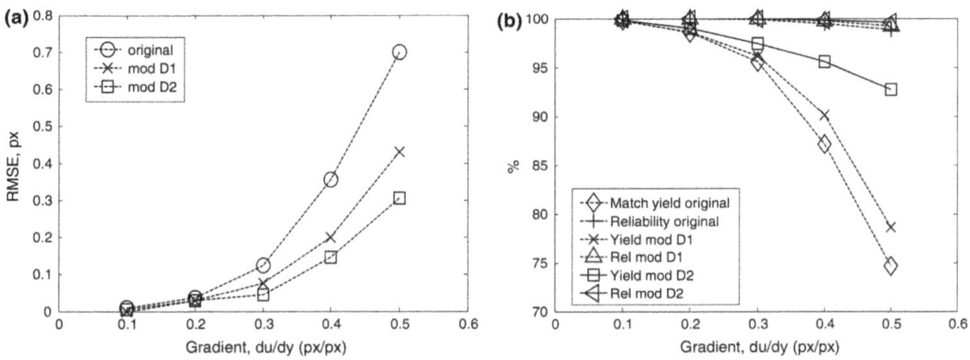

Figure 5.37. Comparisons of the original and modified VB-PTV algorithms with a maximum velocity of 25 pixels. D1 and D2 are the modifications of Paul *et al* [48] that use displacement and directional guidance, respectively. Reprinted from [48] with permission of Springer.

The modification to the proximity matrix has four peaks. One of these corresponds to the displacement calculated using PIV. This means that the algorithm will favor particles with displacement close to the PIV estimate more than particles with zero displacement. The performance improvement of the modification is apparent in figure 5.37, where the modified algorithms achieve both lower RMS error and higher match yield, particularly at high displacements and velocity gradients.

Table 5.13 shows the performace improvement of VB-PTV over previous algorithms. Given the correct particle locations, the reliably identified correct particle tracks at 0.062 N_{ppp}, achieving 0.06 N_{vpp}. Additionally, using the same image, but first performing cascade correlation particle identification (see section 3.2.3) to find 0.032 N_{ppp}, VB-PTV identified 0.027 N_{vpp}. The only comparable results came from the EPTV method.

5.11 Statistical approach

In an effort to develop a computationally efficient, easy-to-implement, and user friendly approach, Fuchs *et al* [52] developed a non-iterative approach to perform tracking for applications in two and three dimensions. This approach is based on

Table 5.13. PTV results with known and unknown particle locations for VB-PTV and previous tracking algorithms using VSJ #301 [47, 51].

Algorithm	Particle location	Matches possible	Matches found	Matches correct	Match yield (%)	Reliability (%)
Present work (tracking only)	Known	4042	4039	3927	97.23	97.15
VAR (Ruhuau *et al* 2005)	Known	4042	4039	3894	96.34	96.41
EPTV (Mikheev and Zubtsov 2008)	Known	4042	3863	3823	94.53	98.96
ICCRM (Brevis *et al* 2011)	Known	4042	NA	3980	98.46	NA
Present work (particle identification + tracking)	Unknown	2095	1846	1761	84.06	95.40
EPTV (Mikheev and Zubtsov 2008)	Unknown	2029	1759	1733	85.41	98.52
VAR (Ruhnau *et al* 2005)	Unknown	NA	872	865	NA	99.20
NRX (Ohmi and Li 2000)	Unknown	NA	808	788	NA	97.52
MF-EPS (Shindler *et al* 2011)	Unknown	NA	1160	1146	NA	98.80
2F-EPS (Shindler *et al* 2011)	Unknown	NA	1123	1112	NA	99.00

three steps. First, all possible particle displacements, within a given range, between a particle, P_1, and its three neighbors in the first image to particles in the second image are obtained. Second, histograms are obtained for all possible displacements of P_1 and its three neighbors, and maxima of these histograms are identified for each direction. Third, the displacement of P_1 to its location in the second image is determined by identifying its smallest absolute deviation from the histograms' maximum values. The VSJ #301 two-dimensional synthetic image set provided by Okamoto *et al* [51] were also used to assess the performance of their tracking algorithm. Here, it was found that it yielded 3846, or 95.15%, valid tracks, with 91, or 2.25% invalid tracks. These results compare well with those shown in table 5.13, coming in third after the vision-based and the VAR methods. In addition, a 3D experimental turbulent boundary layer dataset was also studied. Here, the free stream velocity was 23 m s^{-1}, $Re_\tau = 10\ 000$, and $u_\tau = 0.4$, with a measurement volume of $15 \times 23 \times 0.7$ mm^3, and $N_{ppp} = 0.025$. Using a tomographic predictor for particle locations [53], particle identification and tracking that was effective to $N_{ppp} = 0.0175$ (70% of total particles) were identified (figure 5.38).

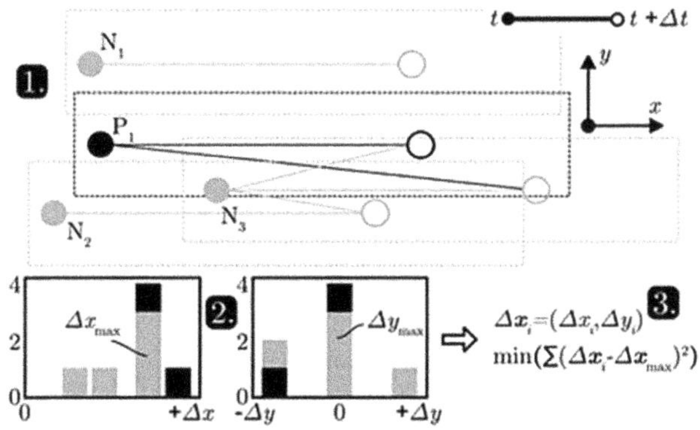

Figure 5.38. (1) Dashed lines: displacement limits; solid lines: possible displacements. (2) Displacement histograms, where the maximum value is determined. (3) Particle P_1 is matched with the black, hollow particle at $t + \Delta t$, since this displacement has the lowest deviation from the histogram maxima [52].

5.12 Outlier detection

In the previously described tracking algorithms, both the yield and the reliability of identified particle tracks are of interest. Algorithms that have a reliability below 100% have at least one incorrect particle track, which not only corrupts the velocity field, but also affects the differential and integral quantities such as vorticity and streamlines [54]. For this reason, it is of the utmost importance to have algorithms determine which particle tracks are outliers and remove them. One technique for outlier detection was developed by Song *et al* [43], which checks that flow results satisfy the continuity equation. The algorithm uses Delaunay tessellation and calculates the divergence within each triangle. In 2D, the flux through each triangle edge is estimated by the mean outward normal component of velocity of the adjacent triangle vertices. The flux is summed for each triangle and compared to a threshold. Spurious vectors are identified as vertices that are part of multiple triangles that exceed a set threshold for divergence. At present, this algorithm is not general due to the incompressibility condition that is imposed.

Jia *et al* [16] and Zhang *et al* [45] both proposed that performing tracking from frame 1–2 and from frame 2–1 can identify spurious vectors. The particle pairs from frame 2–1 are then inverted and combined as shown in figure 5.39. The pairs that have no conflict (5–7) are selected as valid matches. Then if conflicts arise, the match with the most occurrences (1–4 and 3–8) is selected as valid. Conflicts that appear equally (2–6 and 2–9) are considered invalid and removed. This procedure is called either dual computation or bidirectional computing. The results for this method compared to the net-flux method using DT-PTV are shown in figure 5.31, where the bidirectional computation algorithm improved the correct matching rate by roughly 5% for high rates of erased particles.

Sapkota and Ohmi [55] presented an outlier detection method that uses fuzzy logic to determine the probability that a vector is valid. A discussion of fuzzy logic

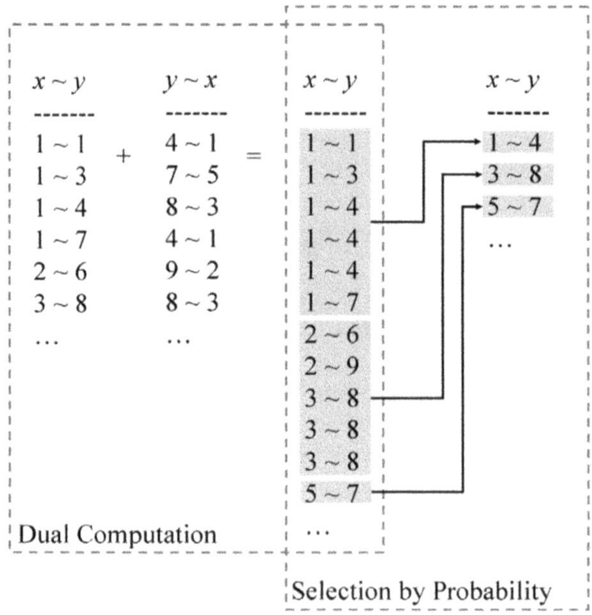

Figure 5.39. Bidirectional computing to remove spurious vectors. Reprinted from [55] with permission.

schemes and the application to identifying particle tracks is discussed in section 5.7.5. For the purpose of identifying spurious vectors, each particle in the first frame is compared to its nearest neighboring particle. From the two corresponding vectors, the fuzzy logic inputs are the distance between vector midpoints, M, the difference in vector magnitudes D, the average value of the vector magnitudes A, and the sum of squares of the differences of the x and y components of the velocity vectors XY. From these inputs, the fuzzy logic system outputs a confidence value, where a high confidence value suggests that a vector is likely valid, while a low confidence suggests that a vector is likely erroneous.

The rule base that determines the confidence outputs corresponding to inputs is shown in table 5.14. As explained in section 5.7.5, the confidence outputs are weighted based on the inputs' degree of membership and summed. This sum is then compared to a threshold value, which was 0.6 in this study. If the confidence exceeds this threshold, the associated vector is considered valid. Invalid vectors were shown to typically have confidence values around 0.1.

The authors used a cross-correlation analysis to determine the range of values for the small, S, medium, M, and large, L, membership ranges for D, A, and XY inputs. Additionally, the range for M inputs was zero to twice the distance between the nearest particles being considered.

This algorithm was tested using a synthetic vortex flow. Figure 5.40 shows that the algorithm successfully detected outliers. It was also suggested that this algorithm can be used to determine the effectiveness of a tracking technique by examining the average and standard deviation values of confidence. A high mean confidence with low standard deviation suggests that a tracking procedure was successful.

Table 5.14. Fuzzy logic rule base for identifying spurious vectors [55].

MS	*DS*			*DM*			*DL*		
	AS	*AM*	*AL*	*AS*	*AM*	*AL*	*AS*	*AM*	*AL*
XYS	High	High	High	High	Med	Med	Med	Med	Low
XYM	High	Med	Low	Med	Med	Med	Low	Low	Low
XYL	Med	Low	Low	Low	Low	Low	Low	Low	Low
MM	*DS*			*DM*			*DL*		
	A5	*AM*	*AL*	*AS*	*AM*	*AL*	*AS*	*AM*	*AL*
XYS	High	Med	Med	Med	Med	Low	Med	Low	Low
XYM	Med	Med	Low	Med	Med	Low	Low	Low	Low
XYL	Low	Low	Low	Low	Low	Low	Low	Low	Low
ML	*DS*			*DM*			*DL*		
	AS	*AM*	*AL*	*AS*	*AM*	*AL*	*AS*	*AM*	*AL*
XYS	Med	Med	Low	Med	Low	Low	Low	Low	Low
XYM	Low	Low	Low	Low	Low	Low	Low	Low	Low
XYL	Low	Low	Low	Low	Low	Low	Low	Low	Low

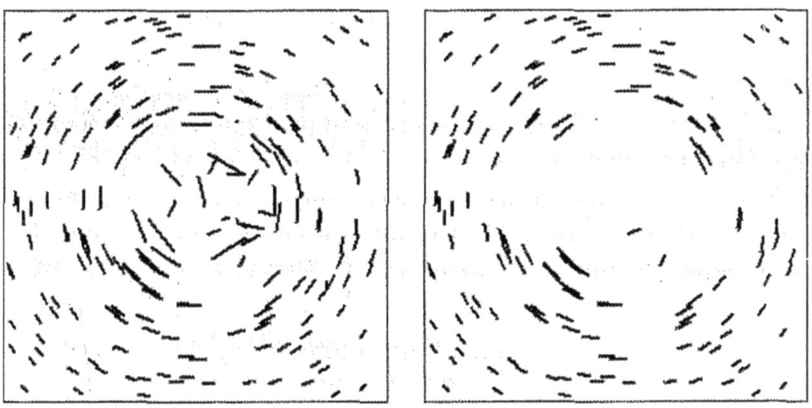

Figure 5.40. Simulated vortex flow with spurious vectors (left) and after spurious vectors have been removed using the fuzzy logic algorithm. Reprinted with permission from [55], copyright ICIC International.

The universal outlier detection method of Westerweel and Scarano [56] is robust for gridded PIV data, and has short computation times. This method has been generalized to work for scattered PTV data by Duncan *et al* [57] In the proposed technique, Delaunay tessellation was used to determine the neighbors of each data point. Neighboring data points share a triangle with one another, and on average each point will have five to eight neighbors using this criterion. These neighbors were used to calculate a normalized residual, which is defined as

Table 5.15. Outlier detection removal rate for PIV and PTV data with added outliers [57].

Added outliers	PIV 1	PIV 2	PTV 1	PTV2
0%	0.10%	0.40%	0.90%	0.60%
5%	5.00%	5.30%	5.70%	5.50%
10%	10.00%	10.00%	10.30%	9.90%
15%	14.40%	16.20%	14.20%	13.80%

$$r_0^* = \frac{|U_0 - \text{med}(U_i)|}{\text{med}|U_i - \text{med}(U_i)| + \varepsilon_a(d + \varepsilon_a)}, \tag{5.63}$$

where U_0 is the velocity measured at the data point in question, U_i is the velocity of each of its neighbors, d is the median of the distances between each of the neighbors and the data point in question, and ε_a is the adaptive tolerance, which was set to 0.1 pixel, as done by Westerweel and Scarano. Table 5.15 shows the performance of the proposed outlier detection scheme for gridded and scattered data. The test was done using synthetic turbulent boundary layer data with added spurious vectors that were randomized with a mean about zero and a maximum equal to the maximum displacement in the flow. The algorithm was able to perform reliably in the presence of up to 15% outliers; however, it must be noted that when outliers are adjacent, the scheme becomes unreliable due to the increase in neighborhood fluctuation. It was suggested that the algorithm performs best when the outlier percentage is at or below 5%.

Masullo and Theunissen [58] noted that clusters of spurious vectors can cause poor performance for the generalized universal outlier detection scheme. Also inspired by a vision-based approach that identifies anomalies based on deviations of magnitudes and angles [59], they suggest using an adaptive generalized universal outlier detection scheme that uses an adaptive neighborhood radius. They argue that for a vector to satisfy a coherency criterion, it must agree with a second-order surface fit to its closest eight neighbors of the form

$$\Phi = a_0 + a_1 y + a_2 x + a_3 y^2 + a_4 xy + a_5 x^2. \tag{5.64}$$

The effect of outliers is minimized by using a diagonal matrix, W, with Gaussian weights

$$W(j+1, j+1) = \exp\left(-1/2\left[\frac{(u_j - u_{\text{median}})^2 + (v_j - v_{\text{median}})^2}{\varepsilon + \sum_j (u_j - u_{\text{median}})^2 + (v_j - v_{\text{median}})^2}\right]^2\right), \tag{5.65}$$

where ε is the background error estimate, and u_{median} and v_{median} are the median values of u_j and v_j, respectively. The parabolic surface fit coefficients are then estimated using a least-squares approach. The coherence value, C is calculated as the residual between the vector in question, (u_0, v_0) and its surface fit, $\Phi_{u,v}$ located at

(x_0, y_0), normalized with the vector magnitudes' median, $|V|_m$, and background error, ε,

$$C = \frac{C_u + C_v}{2},\ C_{(u,v)} = \frac{1}{(|V|_m + \varepsilon)^2}(\Phi_{(u,v)}(x_0, y_0) - (u_0, v_0))^2,\ (|V|_m = \text{median}\left(\left(u_j^2 + v_j^2\right)^{\frac{1}{2}}\right). \tag{5.66}$$

Values of C that exceed the threshold of 10% are considered outliers, which was empirically determined. The coherence is calculated first for the nearest neighbors, and then the neighborhood is progressively increased until at least half of the compared vectors are coherent vectors. This ensures that clusters of incoherent vectors are evaluated in relation to neighbors that comprise of more valid than invalid vectors. Coherence is then recalculated for the increased neighborhood size and compared to the 10% threshold for removal of outliers. This reduces both over-detection and under-detection of spurious vectors.

Their algorithm was further improved by first using an average Gaussian-weighted median estimation and second, by comparing vector direction and magnitude rather than vector components. For the first improvement, the weights, w_i, are obtained using an adaptive Gaussian window (AGW) [60] characterized by an optimal filter width, σ,

$$w_i = \exp\left(\frac{-d_i^2}{\sigma^2}\right), \tag{5.67}$$

with

$$\sigma = \frac{1.24}{N_0}\sum_{i=0}^{N_0} d_i, \tag{5.68}$$

where d_i is the distance between vector i and the vector being analyzed, and N_0 is the number of neighbor particles. The vectors are then organized by increasing magnitude and given widths equal to their weights. This is shown in figure 5.41, where the bar width is given by equation (5.67).The weighted median is shown by the solid black line, and any vector that has a magnitude sufficiently different from this value is considered an outlier. These outliers will also have weights that have extremum values within the vectors' weight set, and in this regard, an averaging interval, Δ_w, centered on the weighted median is defined, where it was found that a value of 0.3–0.5 results in a good compromise between over- and under-detection.

For the second improvement, it was suggested that the normalized residual should be calculated in polar coordinates rather than Cartesian. Thus, a combined residual was defined as

$$r^* = \left(r_\alpha^2 + r_{|V|}^2\right)^{0.5}, \tag{5.69}$$

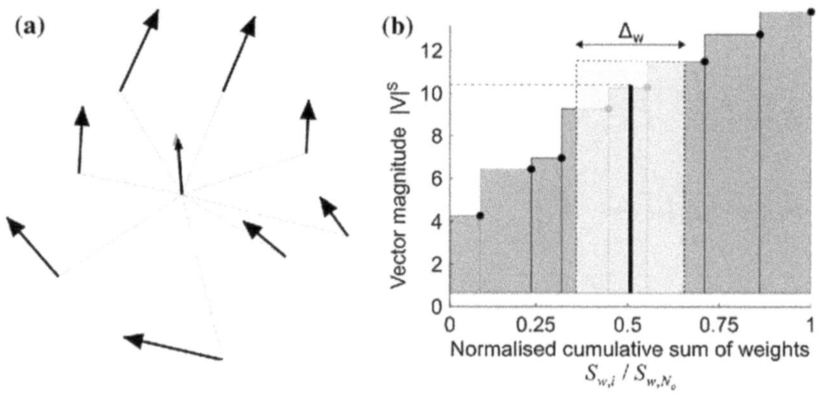

Figure 5.41. (a) Vector validation involves the comparison of the central (black) vector with the predicted vector (grey) on the basis of the neighbouring vectors. The prediction follows from an averageweighted median. (b) Weights in the awmed(•) operator depend on the distance between the investigated vector and its neighbours following (5.67) and are reflected in the bar widths. An interval Δ_w is centred on the weighted median (solid black line), and the weighted average is taken of the data overlapping the defined interval according to (5.68). Reprinted from [58] with permission of Springer.

where the magnitude residual component is

$$r_{|V|} = \frac{||V_0| - |V|_{\text{awm}}|}{\text{median}\left(\left||V_j| - |V|_{\text{awm}}\right|\right) + \varepsilon}, \tag{5.70}$$

$$|V|_{\text{awm}} = \sqrt{\text{awmed}(u_j)^2 + \text{awmed}(v_j)^2}, \tag{5.71}$$

and

$$|V_j|^2 = u_j^2 + v_j^2. \tag{5.72}$$

The parameter ε represents the velocity magnitude background error and is estimated to be 0.1 pixels [56]. The normalized angular residual component is defined similarly as

$$r_\alpha = \frac{\alpha_0^*}{\text{median}\left(\alpha_j^*\right) + \varepsilon_\alpha}, \tag{5.73}$$

where

$$\alpha_j^* = \min\left(\left|\alpha_j - \alpha_{\text{awm}}\right|, 2\pi - \left|\alpha_j - \alpha_{\text{awm}}\right|\right), \tag{5.74}$$

$$\alpha_j = \tan^{-1}\left(\frac{v_j}{u_j}\right), \tag{5.75}$$

$$\alpha_{\mathrm{awm}} = \tan^{-1}\left(\frac{\mathrm{awmed}(\sin(\alpha_j))}{\mathrm{awmed}(\cos(\alpha_j))}\right), \tag{5.76}$$

and

$$\varepsilon_\alpha = \tan^{-1}\left(\frac{\varepsilon}{|V|_{\mathrm{awm}}}\right). \tag{5.77}$$

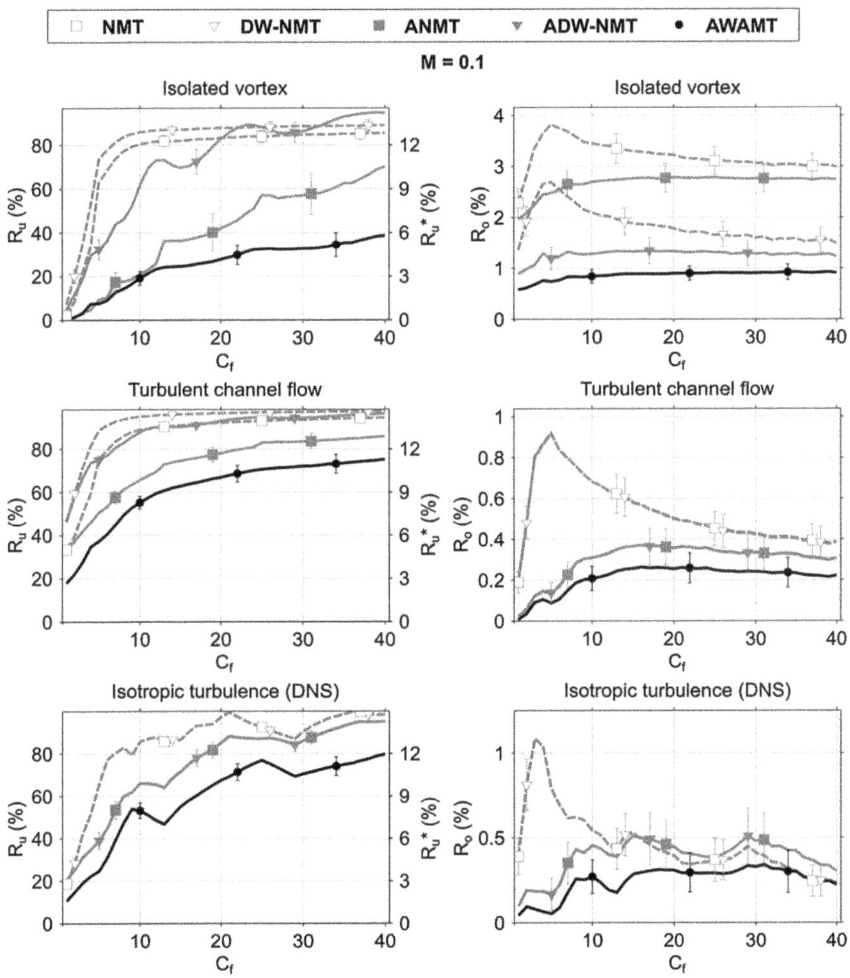

Figure 5.42. Comparison between universal outlier detection (NMT), distance-weighted outlier detection (DW-NMT), NMT and DW-NMT extended with a variable neighbourhood (ANMT and ADW-NMT) and the proposed adaptive weighted angle and magnitude thresholding (AWAMT) in terms of the evolution in under-detection *Ru* and *Ru* * (*left column*) and over-detection *Ro* (*right column*) with increasing cluster size of outliers for the case of unstructured velocity data of a cellular vortex (*top row*), turbulent channel flow (*middle row*) and structured DNS simulation data of isotropic turbulence (*bottom row*). The maximum magnitude of the outlier was set at 10% of the local velocity ($M = 0.1$) [58].

Given these improvements, any vector with a residual r^* below 2 was found to not be valid, which the authors noted agreed with the findings of Westerweel and Scarano [56].

The proposed algorithm was tested for a simulated isolated vortex [61], a turbulent channel flow [62] and a DNS simulation of forced isotropic turbulence generated using DNS [63]. Clusters of spurious points were generated so as to test the ability of the outlier detection schemes in handling large clusters of invalid vectors. The number of vectors included in each cluster, C_f, was varied between 1 and 40, $C_f = 1$ corresponds to individual and non-clustered outliers. The spurious vectors were generated in a random direction and within a magnitude percentage, M, of the mean local velocity. For these tests, M was set to 10% ($M = 0.1$), 100% ($M = 1$), and 1000% ($M = 10$), shown in figure 5.42, figure 5.43, and figure 5.44,

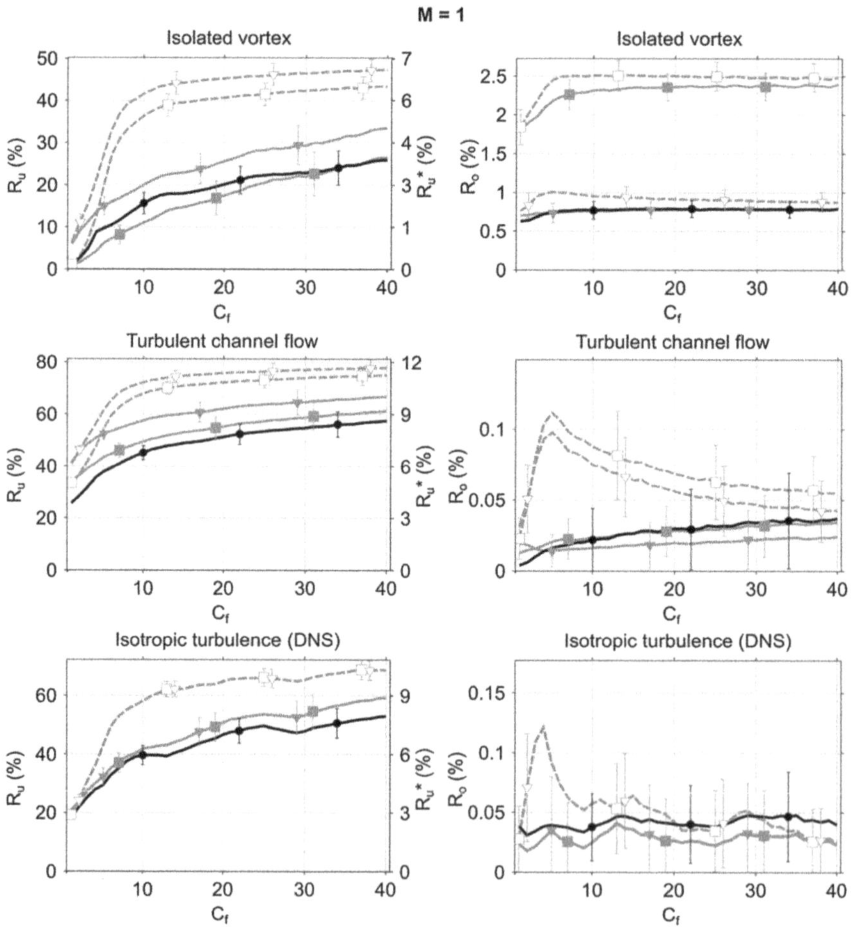

Figure 5.43. Comparison between vector validation methodologies applied to velocity fields contaminated with imposed outlier clusters of varying size and fixed magnitude. The maximum magnitude of the outlier was set at 100% of the local velocity ($M = 1$). See figure 5.42 for further details and legend entries [58].

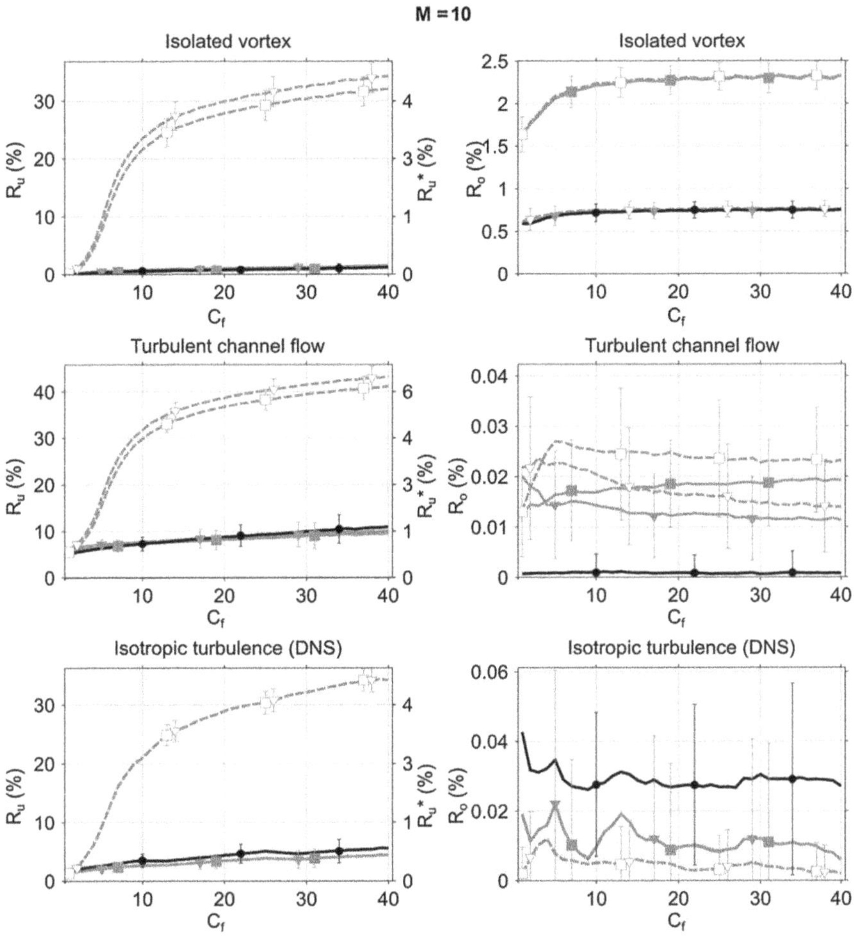

Figure 5.44. Comparison between vector validation methodologies applied to velocity fields contaminated with imposed outlier clusters of varying size and fixed magnitude. The maximum magnitude of the outlier was set at ten times the local velocity ($M = 10$). See figure 5.42 for further details and legend entries [58].

respectively. Several ratios were used to quantify the algorithm's ability to identify spurious vectors: R_u is the ratio of missed outliers to imposed outliers, R_u^* is the ratio of missed outliers to the total number of vectors, and R_o is the ratio of correct vectors that were labeled outliers to the total number of vectors. For the different flow types, it can be seen that the results seem to be dependent on the flow. It can be seen, however, that the variable neighborhood algorithms outperform the generalized universal outlier detection methods, and that the adaptive weighted angle and magnitude thresholding algorithm (AWAMT) outperforms the other algorithms by producing less under- and over-detected outliers. It can also be seen that the under-detected outliers increase with cluster size for the different flow types as well as for the different M values. However, for the over-detected outliers, the behaviors vary with flow type and cluster size. Also seen is that the relative improvements decrease for increasing values of M. In addition, in comparing the different methods, it can be seen that the AWAMT outperforms the other methods for all flow types and M values except perhaps for the isotropic turbulence flow.

The results show that the universal outlier detection algorithm could no longer identify more than 20% of outliers when the cluster size exceeded 10 vectors. Applying only the variable neighborhood improvement increased the outlier detection rate to 80% for 10-vector outlier clusters, although this rate dropped to roughly 30% for outlier clusters of 40 vectors. The proposed AWAMT algorithm succeeded in identifying 60% of outliers when added in clusters of 40 and managed to have a lower over-detection rate and under-detection rate than the compared algorithms for all outlier cluster sizes.

References

[1] Hassan Y and Cannan R 1991 Full-field bubbly flow velocity-measurements using a multiframe particle tracking technique *Exp. Fluids* **12** 49–60

[2] Oullette N, Xu H and Bodenschatz E 2006 A quantitative study of three-dimensional Lagrangian particle tracking algorithms *Exp. Fluids* **40** 301–13

[3] Malik N A, Dracos T and Papantoniou D A 1993 Particle tracking velocimetry in three-dimensional flows—part 2. Particle tracking *Exp Fluids* **15** 279–94

[4] Tarlet D, Bendicks C, Roloff C, Bordas R, Wunderlich B, Michaelis B and Thevenin D 2012 Gas flow measurements by 3D particle tracking velocimetry using coloured tracer particles *Flow Turbul. Combust.* **88** 343–65

[5] Yamamoto F, Uemura T, Tian H Z and Ohmi K 1993 Three-dimensional PTV based on binary cross-correlation method *JSME Int. J.* **36** 279

[6] Hassan Y A, Blanchat T K and Seeley C H 1992 PIV flow visualization using particle tracking techniques *Meas. Sci. Technol.* **3** 633–42

[7] Saga T, Kobayashi T, Segawa S and Hu H 2001 Development and evaluation of an improved correlation based PTV method *J. Visual.* **4** 29–37

[8] Rosenfeld A, Hummel R A and Zcker S W 1976 Scene labeling by relaxation operations *IEEE Trans. SMC* **6** 420–33

[9] Wu Q X and Pairman D 1995 A relaxation labeling technique for computing sea surface velocities from sea surface temperature *IEEE Trans. Geosci. Remote Sens.* **33** 216

[10] Baek S J and Lee S J 1996 A new two-frame particle tracking algorithm using match probability *Exp. Fluids* **22** 23–32

[11] Ohmi K and Li H-Y 2000 Particle-tracking velocimetry with new algorithms *Meas. Sci. Tech.* **11** 603–16

[12] Pereira F, Stüer H, Graff E C and Gharib M 2006 Two-frame 3D particle tracking *Meas. Sci. Technol.* **17** 1680–92

[13] Brevis W, Niño Y and Jirka G H 2011 Integrating cross-correlation and relaxation for particle tracking velocimetry *Exp. Fluids* **50** 135–47

[14] Jia P, Wang Y, Zhang Y and Yang B 2015 Relaxation algorithm-based PTV with dual calculation method and its application in addressing particle saltation *J. Flow Visualization* **18** 71–81

[15] Barnard S T and Thompson W B 1980 Disparity analysis of images *IEEE Trans. Pattern Anal. Machine Intelligence* **2** 333–40

[16] Jia P, Wang Y and Zhang Y 2013 Improvement in the independence of relaxation method-based particle tracking velocimetry *Meas. Sci. Technol.* **24** 055301

[17] Mikheev A V and Zubtsov V M 2008 Enhanced particle-tracking velocimetry (EPTV) with a combined two-component pair-matching algorithm *Meas. Sci. Technol.* **19** 085401

[18] Ruhnau P, Guetter C, Putze T and Schnörr C 2005 A variational approach for particle tracking velocimetry *Meas. Sci. Tech.* **16** 1449–58

[19] Grant I and Pan X 1995 An investigation of the performance of multi layer neural networks applied to the analysis of PIV images *Exp. Fluids* **189** 159–66

[20] Labonté G 1999 A new neural network for particle-tracking velocimetry *Exp. Fluids* **26** 340–6

[21] Ohmi K 2008 SOM-based particle matching algorithm for 3D particle tracking velocimetry *Appl. Math. Comput.* **205** 890–8

[22] Angéniol B, Vaubois G C and Le Texier J Y 1988 Self-organizing feature maps and the traveling salesman problem *Neural Networks* **1** 289–93

[23] Fujimura K, Tokutaka H, Maenou T, Iseki K, Kuwabara E and Ishikawa M 1998 Performance of improved SOM-TSP algorithm for traveling salesman problem of many cities (in Japanese) *IEICE Technical Report: Neuro-Computing* 97–623, pp 95–102

[24] Murai Y, Song X Q, Takagi T, Ishikawa M, Yamamoto F and Ohta J 1999 *Trans. JSME* **65 B** 1339

[25] Ishikawa M, Murai Y, Wada A, Iguchi M, Okamoto K and Yamamoto F 2000 A novel algorithm for particle tracking velocimetry using the velocity gradient tensor *Exp. Fluids* **29** 519–31

[26] Ishikawa M, Murai Y and Yamamoto F 2000 Numerical validation of velocity gradient tensor particle tracking velocimetry for highly deformed flow fields *Meas. Sci. Technol.* **11** 677–84

[27] Ruan X and Zhao W 2005 A novel particle tracking algorithm using polar coordinate system similarity *Acta Mech. Sin.* **21** 430–5

[28] Shindler L, Moroni M and Cenedese A 2010 Spatial–temporal improvements of a two-frame particle-tracking algorithm *Meas. Sci. Technol.* **21** 115401

[29] Stellmacher M and Obermayer K 2000 A new particle tracking algorithm based on deterministic annealing and alternative distance measures *Exp. Fluids* **28** 506–18

[30] Krepki R, Pu Y, Meng H and Obermayer K 2000 A new algorithm for the interrogation of 3D holographic PTV data based on deterministic annealing and expectation minimization optimization *Exp. Fluids* S99–107

[31] Runhau P, Guetter C, Putze T and Schnörr C 2005 A variational approach for particle tracking velocimetry *Meas. Sci. Technol.* **16** 1449–58

[32] Ohyama R and Kaneko K 1997 Experimental study on space and time correspondence of traveling particles for three-dimensional particle image velocimetry by genetic algorithm *Proc. SPIE* **3172** 688–99

[33] Sheng J and Meng H 1998 A genetic algorithm approach for 3D velocity field extraction in holographic particle image velocimetry *Exp. Fluids* **25** 461–73

[34] Doh D H, Kim D H, Cho K R, Cho Y B, Lee W J, Saga T and Kobayashi T 2002 Development of genetic algorithm based 3D-PTV technique *J. Visualization,* **5–3** 243–54

[35] Kimura I, Hattori A and Ueda M 1998 Particle pairing using genetic algorithms for PIV *Proc. VSJ-SPIE* **98** AB-093

[36] Furukawa T, Kimura I, Kuroe Y and Kaga A 1999 Hybrid PTV using neural networks and genetic algorithms *Proc. of the 2nd Pacific Symp. on Flow Visualization and Image Processing* PF-075

[37] Ohmi K and Yoshida N 2000 A new type of GA-based particle tracking velocimetry *Proc. of 9th Int. Symp. on Flow Visualization* #401

[38] Ohmi K and Panday S P 2009 Particle tracking velocimetry using the genetic algorithm *J. Visualization* **3** 217–32

[39] Takagi T 2007 Study on particle tracking velocimetry using ant colony optimization (in Japanese) *J. Visual Soc. Jpn.* **27** 89–90

[40] Ohmi K, Panday S P and Sapkota A 2010 Particle tracking velocimetry with an ant colony optimization algorithm *Exp.Fluids* **48** 589–605

[41] Wernet M P 1993 Fuzzy logic particle tracking velocimetry *Proc. fo the SPIE conference on Optical Diagnostics in Fluid and Thermal Flow (San Diego, CA, 11–16 July)*

[42] Wernet M P 2001 New insights into particle image velocimetry data using fuzzy-logic-based correlation/particle tracking processing *Exp. Fluids* **30** 434–47

[43] Song X, Yamamoto F, Iguchi M and Murai Y 1999 A new tracking algorithm of PIV and removal of spurious vectors using Delaunay tessellation *Exp. Fluids* **26** 371–80

[44] Harrington S 1987 *Computer Graphics—A Programming Approach* (New York: McGraw-Hill) pp 190—5

[45] Zhang Y, Wang Y and Jia P 2014 Improving the Delaunay tessellation particle tracking algorithm in the three-dimensional field *Measurement* **49** 1–14

[46] Zhang Y, Wang Y, Yang B and He W 2015 A particle tracking velocimetry algorithm based on the Voronoi diagram *Meas. Sci. Technol.* **26** 075302

[47] Lei Y-C, Tien W-H, Duncan J, Paul M, Ponchaut N, Mouton C, Dabiri D, Rösgen T and Hove J 2012 A vision-based hybrid particle tracking velocimetry (PTV) technique using a modified cascade-correlation peak-finding method *Exp. Fluids* **53** 1251–68

[48] Paul M, Tien W H and Dabiri D 2014 A displacement-shifted vision-based hybrid particle tracking velocimetry (PTV) technique *Exp. Fluids* **55** 1676

[49] Scott G and Longuet-Higgins H 1991 An algorithm for associating the features of two images *Proc. R Soc. Lond.* B **244** 21–6

[50] Pilu M 1997 A direct method for stereo correspondence based on singular value decomposition *Computer vision and pattern recognition, 1997. Proc., 1997, IEEE Computer Society Conf. on. IEEE* pp 261–6

[51] Okamoto K, Nishio S, Saga T and Kobayashi T Standard images for particle-image velocimetry *Meas. Sci. Tech.* **11** 603–16

[52] Fuchs T, Hain R and Kaehler C J Non-iterative double-frame 2D/3D particle tracking velocimetry *Exp. Fluids* **58** 119

[53] Fuchs T, Hain R and Kähler C J 2016 Double-frame 3D-PTV using a tomographic predictor *Exp. Fluids* **57** 174

[54] Pun C S, Susanto A and Dabiri D 2007 Mode-ratio bootstrapping method for PIV outlier correction *Meas. Sci. Technol.* **18** 3511–22

[55] Sapkota A and Ohmi K 2009 Error detection and performance analysis scheme for particle tracking velocimetry results using fuzzy logic *Int. J. Innovative Comput. Inf. Control* **5** 4927–34

[56] Westerweel J and Scarano F 2005 Universal outlier detection for PIV data *Exp. Fluids* **39** 170–91

[57] Duncan J, Dabiri D, Hove J and Gharib M 2010 Universal outlier detection for particle image velocimetry (PIV) and particle tracking velocimetry (PTV) data *Meas. Sci. Technol.* **21** 057002
[58] Masullo A and Theunissen R 2016 Adaptive vector validation in image velocimetry to minimize the influence of outlier clusters *Exp. Fluids* **57** 33
[59] Reiz B and Pongor S 2011 Psychologically inspired, rule-based outlier detection in noisy data *Proc. 13th Int. Symp. on Symbolic and Numeric Algorithms for Scientific Computing (Timisoara, Romania)*
[60] Aguí J C and Jiménez J 1987 On the performance of particle tracking *J. Fluid Mech.* **185** 447–68
[61] Stanislas M, Okamoto K and Kaehler C J 2001 Main results of the first international PIV challenge *Meas Sci Technol* **14** R63–89
[62] Stanislas M, Okamoto K, Kaehler C J and Westerweel J 2003 Main results of the second international PIV challenge *Exp. Fluids* **39** 170–91
[63] Li Y, Perlman E, Wan M, Yang Y, Burns R, Meneveau C, Burns R, Chen S, Szalay A and Eyink G 2008 A public turbulence database cluster and applications to study Lagrangian evolution of velocity increments in turbulence *J. Turbul.* **9** 31

Chapter 6

Combined tracking and localization for 3D PTV

While standard PTV algorithms identify particle images, determine their spatial coordinates, and then perform tracking, other techniques do not follow this order. The algorithms described in this chapter describe these techniques, which also use aspects of the previously discussed 3D spatial localization (see section 4.2) and particle tracking (see chapter 4) techniques.

Guezennec *et al* [1] first suggested that stereo matching (triangulation with two cameras) could be performed for time-resolved data using particle tracks rather than individual particles. The technique tracks particle images in 2D over five frames (see chapter 4) and matches the tracks seen in each camera using triangulation (see section 4.2.1.2). A cost minimization function quantifies the shortest distance between optical rays for each point in a potential particle track match, as shown in figure 6.1. The stereo matching algorithm was tested up to 0.0034 N_{ppp} and a yield of 98.1% was achieved when exact particle locations were known. When particle locations were identified computationally, the yield decreased to 76.3%.

Another technique was suggested by Kitzhofer and Brücker [2] in order to improve the 3D localization results of their tomo-PTV. They performed a tomographic reconstruction using telecentric lenses (see section 4.2.2) then applied particle tracking using four frame minimum change in acceleration (see section 5.1).

The particle tracks were used to separate real particles from ghost particles. Two criteria were used to identify ghost particle tracks: length and consistency. Short ghost trajectories with tracks that consist of fewer than three consecutive frames were identified and deleted. Longer ghost trajectories with inconsistent trajectories or particle image sizes between frames were also deleted. Both of these types of ghost trajectories are shown in figure 6.2. Notice that the long, real particle track persists over many time steps and the particle image size remains consistent throughout. This technique reduced the erroneous particle tracks from 37% to roughly 5%–10% for a particle density of 0.01 N_{ppp}, which was the highest particle density tested.

doi:10.1088/978-0-7503-2203-4ch6

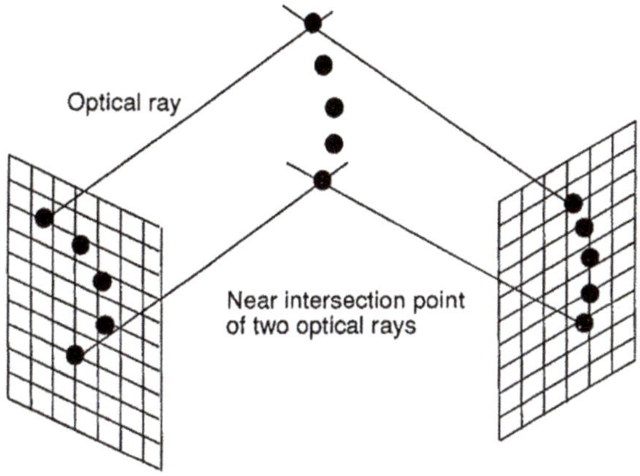

Figure 6.1. Stereo matching for 3D positions of 2D particle tracks. Reprinted from [1] with permission of Springer.

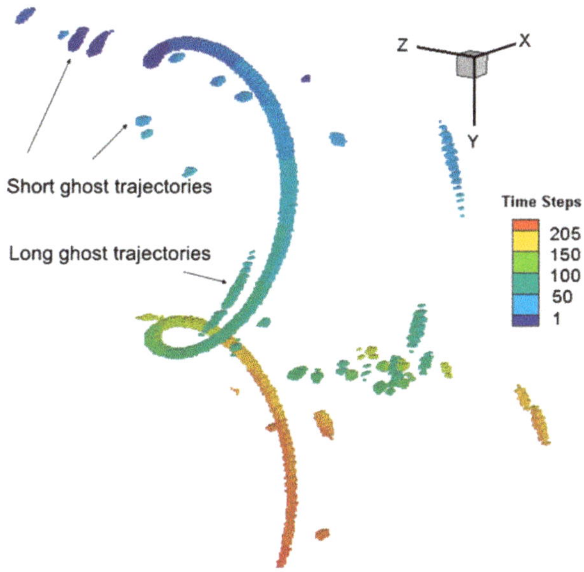

Figure 6.2. Reconstructed trajectories color coded by time step. Short and long ghost trajectories are included. Reprinted from [2] with permission of Springer.

6.1 Shake-the-box

An advanced technique introduced by Schanz *et al* [3–5] is the shake-the-box (STB) method. This method further uses the idea that particle tracks can be used to improve spatial localization results. It relies heavily upon temporal information and tracks from previous time steps to predict particle locations. The predicted particle locations are then 'shaken' in space until they agree with a triangulation. By predicting particle locations, the computation time and number of ghost particles are reduced for high particle seeding densities.

The basic scheme for a single time step using the STB method is as follows:

1. Fit the last n positions of every tracked particle using an optimal Weiner filter.
2. Predict the position of the particle in the next time step $n + 1$ by evaluating the Weiner filter coefficients.
3. 'Shake' the particle predictions iteratively until they line up with triangulation.
4. Find new particles entering the measurement zone using triangulation on the remaining particle images.
5. Shake all particles again to correct for residual errors.
6. Remove particles that are leaving the volume or if the intensity falls below a threshold.
7. Iterate steps 4–6 if necessary.
8. Add new tracks for new particles that are identified within the last four consecutive time steps.

These steps describe how the algorithm handles a time step when nearly all the particle trajectories are known. As none of the trajectories are initially known, there exist three phases of the STB algorithm: initialization, convergence, and converged state.

In the initialization phase, particles are identified using triangulation. Then either tomo-PIV or CFD is used to give an estimate for the trajectory of each particle. Particles that become part of a track that exists for four time steps are sanity checked and added to the system of tracked particles.

Once tracks have been initialized, the Weiner filter, which can be tuned for different signal and noise spectra, extrapolates each track to the next time step. The predicted position is then compared to the images on each camera by calculating and summing the residuals. The projected image is then shifted in the x direction in increments of 0.1 pixels and the position with the smallest residual is selected. The same is then done with the y and z directions. This process is called shaking and is very similar to the technique used in iterative particle reconstruction (section 4.2.1.2).

Shaken particles with an intensity below 5% of the average particle intensity are considered lost and deleted. Further, an intensity correction is applied after each shake iteration to eliminate ghost particles and noise.

At each time step, new particle candidates are identified and new tracks consisting of four time steps are searched. The velocity of particle candidates is predicted by calculating a weighted average of nearby particle tracks. If valid, these tracks are added for predicting the next time step.

In the converged phase, the number of particle tracks should stabilize, as the vast majority of the particles are known and tracked. Convergence can take 20 or more time steps to achieve, depending on seeding density and image noise.

The 3D calibration techniques discussed in section 4.3.3 can all be applied to STB. Schanz *et al* used volumetric self-calibration and the non-uniform optical transfer function. The performance of the technique for noiseless synthetic data is

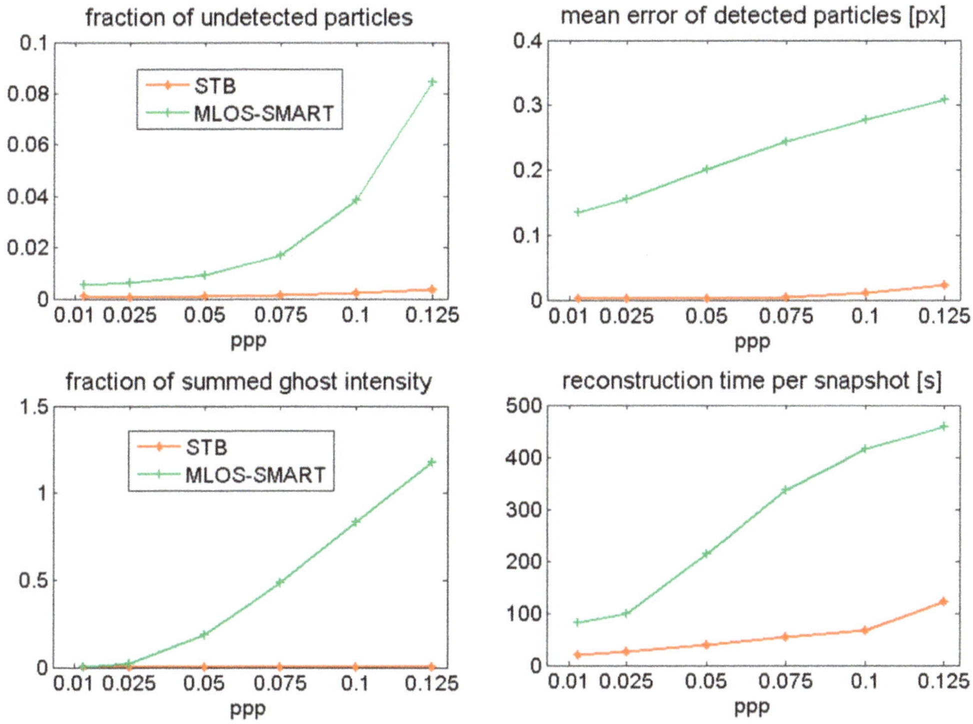

Figure 6.3. Shake-the-box performance versus a MART algorithm. Reprinted from [5] with permission of Springer.

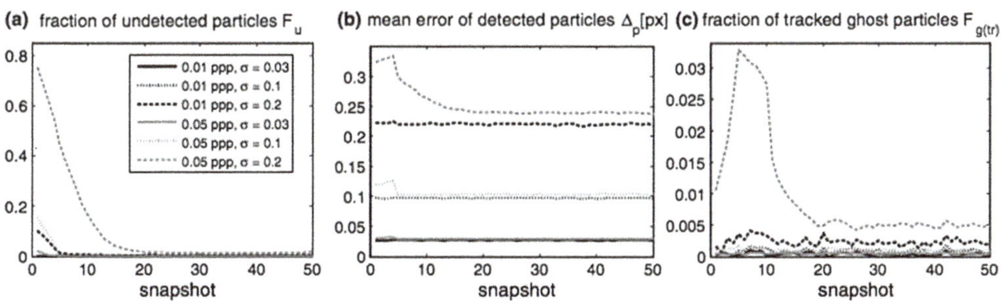

Figure 6.4. Temporal development of STB for 0.01 and 0.05 N_{ppp} with different image noise levels, σ: (a) fraction of undetected particles, (b) particle position accuracy, and (c) fraction of ghost particles. Reprinted from [5] with permission of Springer.

compared to an MLOS-SMART based algorithm in section 4.2.2. It outperforms the MART algorithm in every aspect, particularly for high particle densities (figure 6.3). Figure 6.4 shows the performance of STB for noisy simulated data. The seeding density and noise levels increase the number of time steps needed to reach convergence, the localization error, and the number of tracked ghost particles. In all cases, the STB algorithm achieves more than 95% yield for particle tracks given sufficient time steps and less than 1% tracked ghost particles.

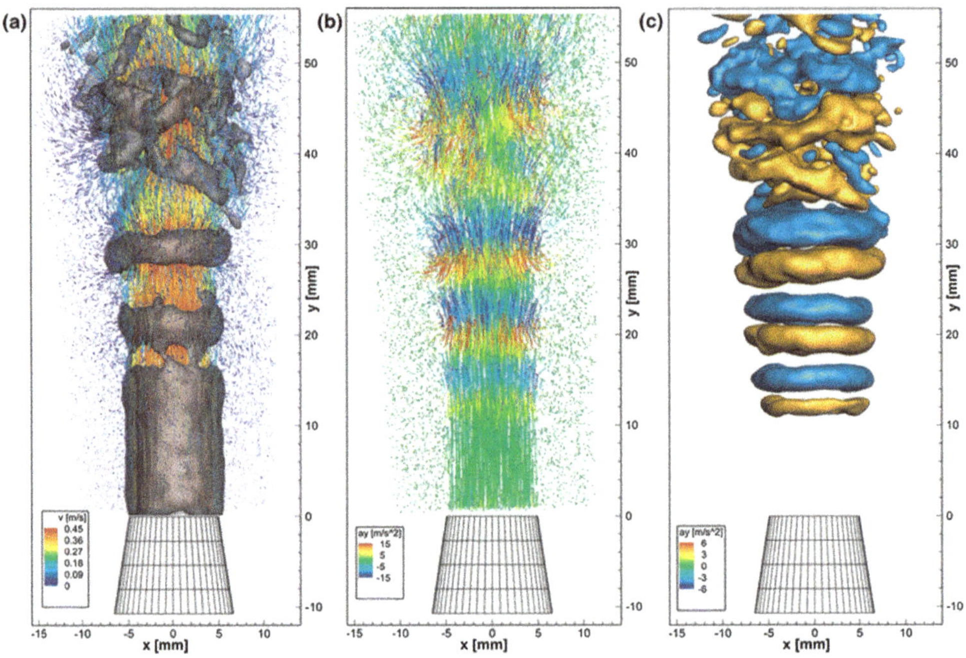

Figure 6.5. (a) Isosurfaces of vorticity (ω = 175/s) as calculated by FlowFit at single time step (t_n) with superimposed particle tracks, extending five time steps back and forth in time; (b) tracked particles for a single time step (t_n, given by spheres) with a tail of ten time steps, color coded by streamwise acceleration; and (c) isosurfaces of streamwise acceleration ($ay = 4$ m s^{-2} and $ay = -4$ m s^{-2}) at a single time step t_n as calculated by FlowFit with penalization of rotation. Reprinted from [5] with permission of Springer.

Shake-the-box was applied to an experiment consisting of a water jet with diameter 10 mm and exit velocity 0.43 m s^{-1}. The average particle image density was 0.035 N_{ppp}. The resulting isosurfaces of vorticity, particle pathlines, and isosurfaces of streamwise acceleration are shown in figure 6.5.

References

[1] Guezennec Y G, Brodkey R S, Trigui N and Kent J C 1994 Algorithms for fully automated three-dimensional particle tracking velocimetry *Exp. Fluids* **17** 209–19

[2] Kitzhofer J and Brücker C 2010 Tomographic particle tracking velocimetry using telecentric imaging *Exp. Fluids* **49** 1307–24

[3] Schanz D, Schröder A, Gesemann S, Michaelis D and Wieneke B 2013 Shake the box: a highly efficient and accurate tomographic particle tracking velocimetry (TOMO-PTV) method using prediction of particle positions *Int. Symp. on Particle Image Velocimetry*

[4] Schanz D, Schröder A and Gesemann S 2014 Shake the box—a 4D PTV algorithm: accurate and ghostless reconstruction of Lagrangian tracks in densely seeded flows *Int. Symp. on Applications of Laser Techniques to Fluid Mechanics*

[5] Schanz D, Gesemann S and Schroder A 2016 Shake-the-box: Lagrangian particle tracking at high particle image densities *Exp. Fluids* **57** 70

IOP Publishing

Dana Dabiri and Charles Pecora

Chapter 7

3D-PTV comparison

A variety of techniques for determining 3D coordinates of tracer particles have been discussed in the previous chapters. The techniques have different strengths and weaknesses, such as the cost of equipment, accuracy, resolution, and volume size. Table 7.1 gives a summary of results obtained for each technique that has been discussed. The summary includes the number of cameras, as well as the resolution and frame rate, used for a particular method. These and the dimensions of the volume give insight into the component requirements and applicability of the different techniques, as some experiments were performed in a microscopic setting, while others are limited to relatively thin volumes. The particle seeding density and resulting vector density are included if the information was available, as these quantities limit the spatial resolution of a 3D algorithm. Note that many experimental applications use ensemble averaging to obtain mean and fluctuating components of velocity over many time steps in order to increase spatial resolution. The root mean square localization error of each technique is also included if the information was available. The accuracy is most often given in terms of pixels, as the camera spatial resolution generally limits accuracy. The tracking technique used for each method is also presented if it was reported. Some techniques were only tested using simulated data as a proof-of-concept; thus, no tracking was performed. The calibration method used in each case is included due to the impact of calibration on the accuracy of an experiment. The calibration techniques are discussed in detail in section 4.3.

doi:10.1088/978-0-7503-2203-4ch7

Table 7.1. Comparison of 3D-PTV techniques.

Authors	N_{ppp}	N_{vpp}	In-plane error	Out-of-plane error	Volume size (mm^3)	# of cameras	Camera resolution	Frame rate	Calibration	Notes	Tracking algorithm
Astigmatism, stereoscopic (experiment) Fuchs *et al* (2014) [1]	5.00E-04	5.00E-04	0.01% of depth (0.13 px)	0.03% of depth (0.39 px)	40 × 40 × 20	2	2560 × 2160	NR	Depth curve was fitted to distorted particle images		NR
Astigmatism micro PTV (experiment) Ichikawa *et al* (2017) [2]	NR	NR	NR	NR	0.348 × 0.261 × 0.060	1	960 × 720	45 Hz	Depth curve was fitted to distorted particle images		NR
Defocusing (experiment) Fuchs *et al* (2016) [3]	1.00E-04	NR	0.06 px	0.64 px	20 × 20 × 1	1	NR	NR	Volumetric self-calibration	Thin wall-bounded measurement volume	Nearest neighbor
Defocusing micro PIV (experiment) Tien *et al* (2014) [4]	NR	4.20E-04	0.18 px	1.83 px	3.35 × 2.5 × 1.5	1	1024 × 1024	NR	RBF model-free calibration	Resulted in 0.0004 N_{vpp}	Vision-based tracking
Iterative particle techniques, reconstruction (noiseless simulation) Wieneke (2013) [5]	0.0004-0.2	NR	<0.1 px (for <0.05 N_{ppp})		1000 × 1000 × 300 vx	4	1300 × 1300	NR	Volumetric self-calibration		NR
Iterative particle reconstruction (experiment) Wieneke (2013) [5]	3.00E-02	NR	NR	NR	50 × 30 × 30 (cylindrical)	4	1024 × 1024	1000 Hz	Volumetric self-calibration	Water jet	NR

Plenoptic (experiment) Chen and Sick (2017) [6]	NR	3.00E-06	NR	NR	24 × 22 × 6	1	29 megapixels	1 Hz	NR	Reported 0.21 particles per micro-lens, which was not described in terms of pixels. Data was time-averaged for 400 image pairs.	Nearest neighbor
Scanning PTV (experiment) Hoyer *et al* (2005) [7]	4.70E-03	3.30E-03	NR	NR	40 × 40 × 20	1	1024 × 1024	50 Hz	Ray tracing	A four-view image splitter was used. The camera recorded at 500 Hz to obtain volumetric scans at 50 Hz.	Minimum acceleration
Shake-the-box (experiment) Novara *et al* (2016) [8]	3.50E-02	2.80E-02	0.1 px (estimate for synthetic images)		50 × 90 × 8	8	2560 × 2160	10 Hz (four frames per recording)	Optical transfer function and volume self-calibration	Four time steps are captured using two separate, four-camera imaging systems and two polarized lasers.	Multi-frame tracking with a tomo-PIV predictor.
Shake-the-box (experiment) Schanz *et al* (2016) [9]	3.50E-02	1.64E-02	NR	NR	55 × 25 × 25 (cylindrical)	4	672 × 1024	1000 Hz	Optical transfer function and volume self-calibration	Water jet	STB

(*Continued*)

Table 7.1. (*Continued*)

Authors	N_{ppp}	N_{vpp}	In-plane error	Out-of-plane error	Volume size (mm^3)	# of cameras	Camera resolution	Frame rate	Calibration	Notes	Tracking algorithm
Shake-the-box (noiseless simulation) Schanz *et al* (2016) [9]	1.25E-01	1.25E-01	0.018 px		1000 × 1000 × 400 vx	4	1200 × 1200	NR	Optical transfer function and volume self-calibration		STB
Stereoscopic (noiseless simulation) Guezennec *et al* (1994) [10]	3.40E-03	2.60E-03	1.0 px	1.0 px	NR	2	512 × 512	NR	NR		Minimum change in acceleration
Stereoscopic ant colony optimization (PIV standard images) Panday *et al* (2011) [11]	0.0024–0.024	NR	NR	NR	NR	2	256 × 256	NR	NR	Yield was 72% for 0.024 N_{ppp} and 93% for 0.0024 N_{ppp}	NR
Tomographic-PTV (experiment) Schneiders and Scarano (2016) [12]	7.00E-02	NR	NR	NR	11 × 5 × 24	4	NR	10 000 Hz	NR		Minimum change in acceleration
Tomographic-PTV (experiment) Fuchs *et al* (2016) [3]	4.00E-03	NR	0.125 px	0.15 px	20 × 20 × 1	4	NR	NR	Volumetric self-calibration	Thin wall-bounded measurement volume	NR
Tomographic-PTV (simulation) Doh *et al* (2012) [13]	3.80E-02	NR	0.5 vx	0.5 vx	35 × 35 × 8	4	512 × 512	NR	Ray tracing		Relaxation
Triangulation (experiment) Maas *et al* (1993) [14]	3.80E-03	3.20E-03	0.15 px	0.46 px	200 × 160 × 50	3	512 × 512	30 Hz	Ray tracing		Minimum change in acceleration

Triangulation (experiment) Shirsath *et al* (2015) [15]	3.00E-03	NR	0.5 mm	1.0 mm	NR	3	128 × 1024	2000 Hz	Ray tracing	Granular flow down a rotating chute	Minimum change in acceleration
Triangulation (experiment) Alberini *et al* (2017) [16]	NR	NR	NR	NR	NR	1	1024 × 1024	1000 Hz	NR	Algorithms were obtained from Photrak AG (Hoyer *et al* 2005) [7]	Tracking algorithm obtained from Wilneff (2002) [23]
Triangulation (experiment) Kim *et al* (2016) [17]	3.30E-04	NR	NR	NR	4 × 4 × 4	1	1728 × 1728	550 Hz	Ray tracing		Minimum change in acceleration
Triangulation (experiment) Rochlitz *et al* (2015) [18]	NR	NR	0.13% of max disp	0.3 px	NR	2	NR	NR	Volumetric self-calibration	Cameras were oriented at 90° from one another	Nearest neighbor
Triangulation (experiment) Janke *et al* (2017) [24]	1.00E-03	NR	2.1 px	2.1 px	NR	3	NR	500 Hz	Ray tracing		Minimum change in acceleration
Triangulation guided by tomographic reconstruction (noiseless simulation) Fuchs *et al* (2016) [19]	5.00E-02	NR	0.1 px	0.15 px	8 × 5 × 2.5	4	NR	10 200 Hz	Volumetric self-calibration	Effective N_{ppp} was 0.035 and the SNR tested was 106	NR
DIH-PTV (experiment) Toloui *et al* (2017) [20]	NR	3.50E-03	?	?	10 × 50 × 10	1	1024 × 1024	3000 Hz	None		Iterative multi-frame tracking algorithm

(*Continued*)

Table 7.1. (*Continued*)

Authors	N_{ppp}	N_{vpp}	In-plane error	Out-of-plane error	Volume size (mm^3)	# of cameras	Camera resolution	Frame rate	Calibration	Notes	Tracking algorithm
DIH-PTV (noiseless simulation) Toloui and Hong (2015) [21]	2.40E-03	NR	3.0 px	3.0 px	2.05 × 2.05 × 0.5	1	2048 × 2048	NR	None		NR
DIH-PTV stereoscopic (experiment) Buchmann *et al* (2013) [22]	NR	NR	0.23 px	0.23 px	3 × 3 × 5	2	312 × 260	500 000 Hz	Linear transform		NR

References

[1] Fuchs T, Hain R and Kähler C J 2014 Macroscopic three-dimensional particle location using stereoscopic imaging and astigmatic aberrations *Opt. Lett.* **39** 6863–6

[2] Ichikawa Y, Yamamoto K, Yamamoto M and Motosuke M 2017 Near-hydrophobic-surface flow measurement by micro-3D PTV for evaluation of drag reduction *Phys. Fluids* **29** 092005

[3] Fuchs T, Hain R and Kähler C J 2016 *In situ* calibrated defocusing PTV for wall-bounded measurement volumes *Meas. Sci. Technol.* **27** 084005

[4] Tien W H, Dabiri D and Hove J R 2014 Color-coded three-dimensional micro particle tracking velocimetry and application to micro backward-facing step flows *Exp. Fluids* **55** 1684

[5] Wieneke B 2013 Iterative reconstruction of volumetric particle distribution *Meas. Sci. Technol.* **24** 024008

[6] Chen H and Sick V 2017 Three-dimensional three-component air flow visualization in a steady-state engine flow bench using a plenoptic camera *SAE Int. J. Engines* **10** 625–35

[7] Hoyer K, Holzner M, Lüthi B, Guala M, Liberzon A and Kinzelbach W 2005 3D scanning particle tracking velocimetry *Exp. Fluids* **39** 923–34

[8] Schanz D, Reuther N, Kaehler C, Schroeder A and Novara M 2016 Lagrangian 3D particle tracking in high-speed flows: shake-the-box for multi-pulse systems *Exp. Fluids* **57** 128

[9] Schanz D, Gesemann S and Schroder A 2016 Shake-the-box: Lagrangian particle tracking at high particle image densities *Exp. Fluids* **57** 70

[10] Guezennec Y, Brodkey R, Trigui N and Kent J 1994 Algorithms for fully automated three-dimensional particle tracking velocimetry *Exp. Fluids* **17** 209–19

[11] Panday S P, Ohmi K and Nose K 2011 An ant colony optimization based stereoscopic particle pairing algorithm for three-dimensional particle tracking velocimetry *Flow Meas. Instrum.* **22** 86–95

[12] Schneiders J F and Scarano F 2016 Dense velocity reconstruction from tomographic PTV with material derivatives *Exp. Fluids* **57** 139

[13] Doh D H, Cho G R and Kim Y H 2012 Development of a tomographic PTV *J. Mech. Sci. Technol.* **26** 3811–19

[14] Maas H G, Gruen A and Papantoniou D 1993 Particle tracking velocimetry in three-dimensional flows Part 1. Photogrammetric determination of particle coordinates *Exp. Fluids* **15** 133–46

[15] Shirsath S S, Padding J T, Clercx H J H and Kuipers J A M 2015 Cross-validation of 3D particle tracking velocimetry for the study of granular flows down rotating chutes *Chem. Eng. Sci.* **134** 312–23

[16] Alberini F, Liu L, Stitt E H and Simmons M J H 2017 Comparison between 3-D-PTV and 2-D-PIV for determination of hydrodynamics of complex fluids in a stirred vessel *Chem. Eng. Sci.* **171** 189–203

[17] Kim J-T, Kim D, Liberzon A and Chamorro L P 2016 Three-dimensional particle tracking velocimetry for turbulence applications: case of a jet flow *J. Vis. Exp.* **108** e53745

[18] Rochlitz H, Scholz P and Fuchs T 2015 The flow field in a high aspect ratio cooling duct with and without one heated wall *Exp. Fluids* **56** 208

[19] Fuchs T, Hain R and Kähler C J 2016 Double-frame 3D-PTV using a tomographic predictor *Exp. Fluids* **57** 174

[20] Toloui M, Mallery K and Hong J 2017 Improvements on digital inline holographic PTV for 3D wall-bounded turbulent flow measurements *Meas. Sci. Technol.* **28** 044009
[21] Toloui M and Hong J 2015 High fidelity digital inline holographic method for 3D flow measurements *Opt. Express* **23** 10.1364
[22] Buchmann N A, Atkinson C and Soria J 2013 Ultra-high-speed tomographic digital holographic velocimetry in supersonic particle-laden jet flows *Meas. Sci. Technol.* **24** 024005
[23] Willneff J 2002 3-D particle tracking velocimetry based on image and object space information *ISPRS Commission V Symp. (Corfu, Greece, September 2–6, 2002)*
[24] Janke T, Schwarze R and Bauer K 2017 Measuring three-dimensional flow structures in the conductive airways using 3D-PTV *Exp. Fluids* **58** 133

IOP Publishing

Particle Tracking Velocimetry

Dana Dabiri and Charles Pecora

Chapter 8

Post-processing

The primary goal of PTV algorithms is to provide velocity estimates at discrete, scattered points within a measurement space. In many cases, differential quantities, such as vorticity, strain, and acceleration, are of interest. As PIV data is generated on a grid, velocity gradients can be estimated using finite difference methods, which are shown in figure 8.1. The second-order central difference scheme predicts velocity derivatives for gridded data using

$$\frac{\partial u_i}{\partial x_j}(\boldsymbol{x}_k) = \frac{u_i(\boldsymbol{x}_{k+1}) - u_i(\boldsymbol{x}_{k-1})}{2\delta x_j}, \tag{8.1}$$

where $\boldsymbol{x}_k$ is the point at which the velocity derivative is being estimated, and $\boldsymbol{x}_{k+1}$ and $\boldsymbol{x}_{k-1}$ are the points located $+\delta x_j$ and $-\delta x_j$ from point $\boldsymbol{x}_k$, respectively [1]. The central difference scheme can also be extended to a fourth-order estimate [1],

$$\frac{\partial u_i}{\partial x_j}(\boldsymbol{x}_k) = \frac{u_i(\boldsymbol{x}_{k+2}) + 8u_i(\boldsymbol{x}_{k+1}) - 8u_i(\boldsymbol{x}_{k-1}) - u_i(\boldsymbol{x}_{k-2})}{12\delta x_j}. \tag{8.2}$$

For simply calculating the vorticity at a point, the eight-point circulation method can also be used. This is done by calculating the circulation around point $\boldsymbol{x}_k$ using the tangential components of velocity at neighboring points [2].

These methods of estimating spatial velocity derivatives are used with regularly gridded data, thus PTV vectors are typically interpolated for analysis. Interpolation schemes and their use for calculating differential quantities will be discussed in section 8.1.

Temporal velocity derivatives can be estimated when velocity measurements are time-resolved. For Lagrangian particle tracks, the acceleration of a tracer particle over a single time step gives the material derivative

$$\frac{D\boldsymbol{u}}{Dt} = \frac{\partial \boldsymbol{u}}{\partial t} + (\boldsymbol{u} \cdot \nabla)\boldsymbol{u}, \tag{8.3}$$

doi:10.1088/978-0-7503-2203-4ch8

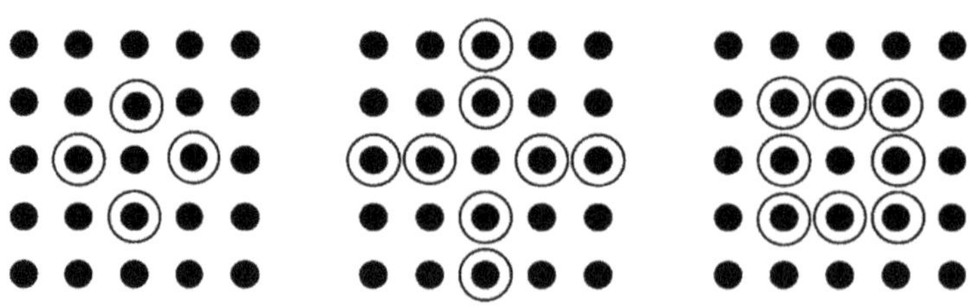

Figure 8.1. Velocity measurement locations used for estimation of spatial derivatives using (a) second-order central difference, (b) fourth-order central difference, and (c) eight-point circulation methods. Reprinted from [2] with permission of Springer.

where the acceleration can be integrated towards obtaining the pressure field within a flow. Pressure calculation from PTV data will be discussed in section 8.2.

8.1 Interpolation

One of the first and most common methods for interpolating randomly distributed particle tracking vectors is adaptive Gaussian window interpolation (AGW) [3]. The proposed method generates velocity estimates on a grid, where it was suggested to use a grid spacing, H, equal to 1.24 times the mean particle spacing. The velocity estimate at point x is then given as

$$u(x) = \frac{\sum_i \alpha_i u_i}{\sum_i \alpha_i}, \tag{8.4}$$

where u_i are the known velocities measured at positions x_i and

$$\alpha_i = \exp\left(\frac{-|x - x_i|^2}{H^2}\right). \tag{8.5}$$

Agui and Jimenez [3] also proposed a constrained smoothing algorithm. First, an estimate of the interpolation error is obtained using a bootstrapping method. Bootstrapping is performed by taking many randomly selected subsets of the given data and estimating the remaining data points using interpolation. These results are statistically analyzed to obtain an estimate of the interpolation error. Second, this error estimate is then used as a constraint for the smoothing step; the smoothing is done by iteratively replacing each interpolated value with the mean of its four neighbors if it exists within the appropriate one standard deviation error bound. Otherwise, the error bound is selected as the smoothed value. Iterating this formula was shown to quickly converge to a solution.

After interpolating and smoothing, a second-order central difference scheme is used to estimate the velocity derivatives. Figure 8.2 shows the vorticity RMS error and bootstrap estimated variance before and after smoothing versus h/δ, where h is the distance between grid points used for the central difference scheme, and δ is the

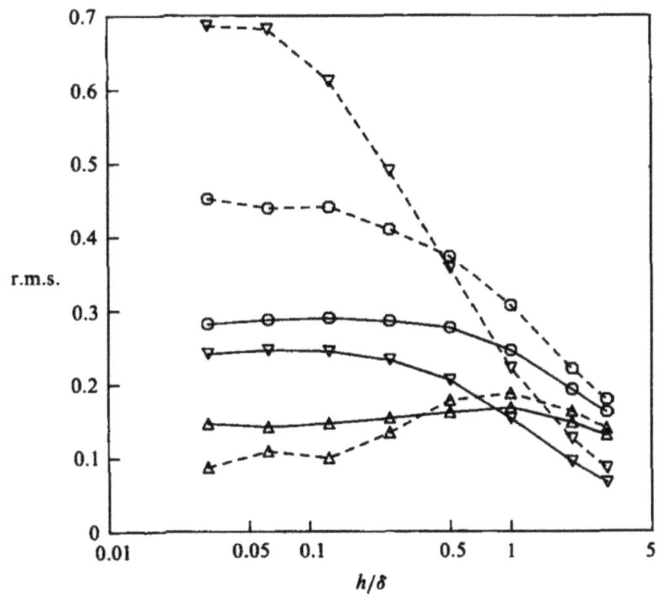

Figure 8.2. RMS of vorticity estimation on interpolation parameters, H/δ and h/δ. Unsmoothed vorticity (circle); bootstrap predicted variance (∇); smoothed vorticity (Δ). $H/\delta = 1.24$ (solid line); $H/\delta = 0.62$ (dashed line). From [3], reproduced with permission.

mean particle spacing. The high frequency fluctuations, which appear at small differentiation distances, are smoothed out when a large interpolation window, H, is used. When the smoothing process is applied, the vorticity RMS is reduced across all ranges of h/δ to roughly the same level regardless of interpolation window size.

It was later suggested that least-squares Taylor polynomials could be used to interpolate data and the coefficients could be directly used to determine differential quantities [4]. Cohn and Koochesfahani [2] tested the effects of vector density and noise on this differentiation scheme. The grid size was selected such that the vector density of the remapped field is the same as the initial irregularly spaced field. Tests were done to compare second, third, and fourth-order Taylor polynomial fits with a fit radius R for interpolation of a simulated Oseen vortex with a grid spacing of $L/\delta = 3.0$, where L is the vortex core radius. Figure 8.3 shows that the most substantial interpolation error is present as a bias within the vortex core. This bias is a maximum for the second- and third-order Taylor polynomials with large fit radii. The minimum bias error was found using a fourth-order fit with $R/\delta = 3$, although its advantage was minimal compared to a second-order fit with $R/\delta = 2$. The random component of error was small compared to the bias error, and larger radius fits performed better with more noise and worse with less noise; therefore, the authors suggested using a second-order fit with $R/\delta = 2$ for simplicity.

Using this interpolation method, Cohn and Koochesfahani [2] compared the previously described second-order central difference, fourth-order central difference, and eight-point circulation methods with a direct differentiation of a fourth-order

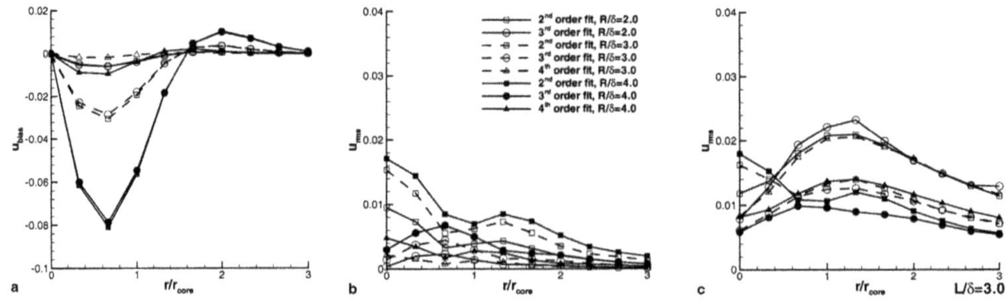

Figure 8.3. Accuracy of interpolated velocity field as (a) bias error with 0% noise, (b) random error with 0% noise, and (c) random error with 6% noise. Reprinted from [2] with permission of Springer.

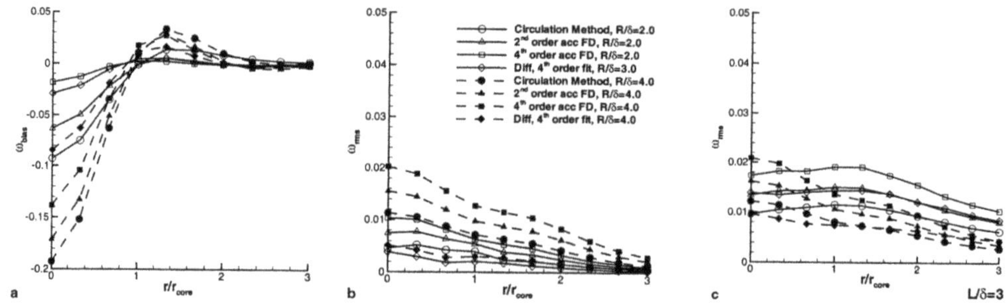

Figure 8.4. Accuracy of out-of-plane vorticity field computed for different values of R/δ using four different differentiation methods: (a) bias error, (b) random error with 0% added noise, and (c) random error with 6% added noise. Reprinted from [2] with permission of Springer.

interpolation polynomial. The same Oseen vortex flow used for the interpolation comparison was also used for this analysis, where its results are shown in figure 8.4.

The fourth-order central difference and polynomial differentiation using the smallest interpolation radius yielded the lowest bias error, however, these also gave the greatest random errors in the presence of noise. Again, the difference in bias error is more substantial than the difference of RMS, so either the fourth-order finite difference method or the polynomial differentiation technique was considered best. For the sake of simplicity and computational efficiency, the authors suggested using the fourth-order finite difference method, as both the bias and random components of error were less than 2% of the maximum vorticity in the presence of 6% noise.

Ido *et al* [5] proposed solving ellipsoidal equations to interpolate data points onto a regular grid and iteratively correct them according to the continuity equation. The first technique solves the Laplace (LER) equation

$$\nabla^2 \boldsymbol{u} = 0, \tag{8.6}$$

with PTV vectors as fixed points, which results in a linear interpolation onto any desired grid. Using this Laplace equation, the authors found that the first derivatives of the interpolated velocity field are not smooth. In order to have smooth derivatives, the authors suggested a second technique that uses the biquadratic-ellipsoidal (BER) equation

$$\nabla^4 \boldsymbol{u} = 0, \tag{8.7}$$

which gives third-order spatial continuity, yielding smooth derivative estimates. A velocity correction potential (VCP) was also introduced to iteratively adjust interpolated vectors to satisfy the continuity equation. An improve velocity correction potential (IVCP) was also proposed, in which the measured PTV vectors can be adjusted in order to improve spatial smoothness and reduce the impact of noise.

The LER, BER, inverse distance rearrangement (IDR), and AGW interpolation methods were tested using a Taylor–Green vortex flow field. LER and BER were tested both with and without the smoothing corrections. Figure 8.5 shows the correlation coefficient of vorticity and the relative error for vorticity, which is simply the vorticity error normalized by the largest vorticity in the field, versus the vector density. The performance of all algorithms improves with an increase in the number of vectors per vortex. The authors therefore stated that the number of vectors per vortex should be selected so that normalized vorticity error below 10% is achieved as a good measure of the performance of an algorithm. LER, IDR, and AGW all required at least 50 vectors per vortex to achieve the desired error level, while BER required only 14. With the addition of IVCP smoothing, the BER algorithm achieved less than 10% error with only 11 vectors per vortex.

In a later paper, Ido and Murai [6] described a technique for interpolation using the hexagonal-ellipsoidal equation rearrangement (HER),

$$\nabla^6 \boldsymbol{u} = 0. \tag{8.8}$$

In order to obtain a unique solution for the sixth-order differential equation, the LER and BER interpolations need to be performed and used as a computational initial condition. It was shown that the HER algorithm was capable of reconstructing a more valid energy spectrum than the BER and LER algorithms in the case of isotropic turbulence.

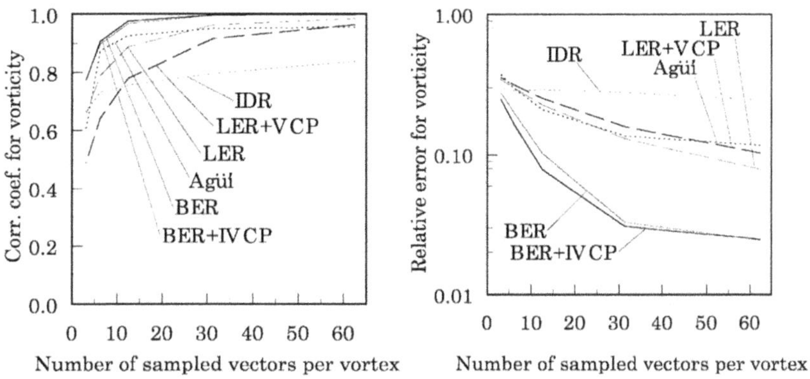

Figure 8.5. Plot of the vorticity correlation coefficient (left) and normalized vorticity error (right) of the proposed ellipsoidal equation algorithms with and without smoothing. Reprinted from [5] with permission of Springer.

Vedula and Adrian [7] suggested using a linear stochastic estimator for interpolating velocity onto a grid. It was suggested that an optimal estimate will have the minimum mean square error when compared to the measured point velocities. This is done by defining a Cartesian component of the filtered velocity field as

$$\widetilde{u}_i(\boldsymbol{x}, t) = \int h_{ij}(\xi, \boldsymbol{x}) u_j(\xi, t) d\xi, \tag{8.9}$$

where $\boldsymbol{x}$ is the position vector, h_{ij} denotes the filter impulse response function, and u_j is a Cartesian component of the unfiltered (measured) velocity field. In the case of incompressible flow, the solenoidal property was imposed while finding the optimal filter, so that the divergence of the filtered velocity field is zero. By expanding the filtered estimate to a first-order power series, the velocity and its spatial derivatives are estimated such that the divergence is zero and the mean square error of the estimate with respect to the measured point velocities is a minimum.

In order to evaluate the performance of this interpolation algorithm, DNS was used to generate incompressible, isotropic turbulence on a 256 [3] grid. The domain was subdivided into non-overlapping cubic sub-domains, in which the velocity was sampled at M scattered points. A non-dimensional data density was defined as $\psi \equiv M\lambda^3/L_s^3$, where λ is the Taylor micro-scale of the flow and L_s is the side length of each cubic sub-domain. Additionally, a Fourier cut-off filter of width $\Delta_c \equiv \pi/k_c$, where k_c is the cut-off wavenumber, was applied to the simulated data. The resulting mean squared error from this optimal interpolator, e, was normalized by the variance of the filtered velocity field, $\sigma_{\widetilde{u}}^2$ and is shown as a function of filter width in figure 8.6. For comparison using the highest normalized data density, the normalized error was also determined using AGW interpolation, shown by the dotted line. Applying the optimal filter width for this data density resulted in a normalized error of 4% while AGW resulted in 14% error. The filter width

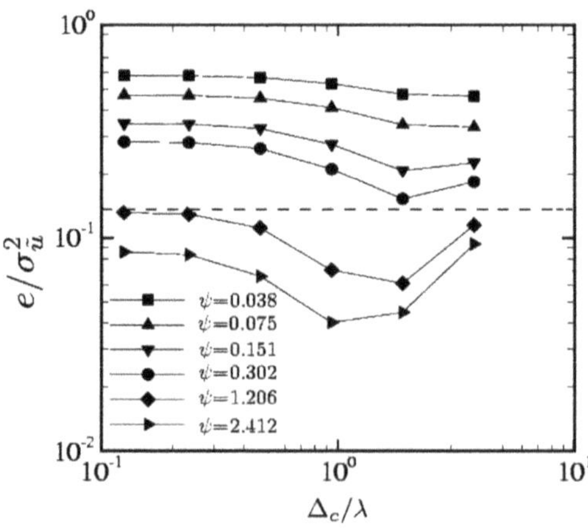

Figure 8.6. Noramlized mean square error versus normalized cut-off filter widths at various data densities, ψ, and a sub-volume size of $L_s/\lambda = 3.756$. The dashed line shows the error for AGW for $\psi = 2.412$. Reprinted from [7] with permission of Springer.

corresponding to a minimum normalized error appears to decrease with increase in data density.

Casa and Krueger [8] proposed using Gaussian weighted radial basis functions (RBF) for interpolation. An advantage of the RBF interpolation is that it is infinitely smooth and can be used to find smooth velocity gradients through differentiation. Additionally, boundary conditions and constraints, such as incompressibility, can be introduced. The general form of Gaussian RBFs is

$$u_j(\boldsymbol{x}) = \sum_{i=1}^{N} b_{ji}\exp\left[\frac{-\|\boldsymbol{x} - \boldsymbol{\mu}_i\|^2}{2\sigma_{ji}^2}\right], \tag{8.10}$$

where u_j is the function approximation, $\boldsymbol{x}$ is the position vector, N is the number of RBFs used, b_{ij} are RBF weights, $\boldsymbol{\mu}_i$ are the locations of the RBF centers, and σ_{ji} are the widths of the Gaussian RBFs. The $\boldsymbol{\mu}_i$ locations were fixed within the fluid volume while b_{ij} and σ_{ji} were determined using nonlinear least-squares optimization. The two parameters were optimized independently in order to reduce computation time.

The interpolation algorithm was applied to an experimental 3D dataset of an unbounded vortex ring. Both the AGW and RBF interpolation methods were used to identify the vortex center, axis, and radius. The AGW vorticity estimates were obtained using central differences, while the RBF estimates were found via differentiation of $u_j(\boldsymbol{x})$. The impact of noise on the AGW determination of vorticity is shown in figure 8.7. Both methods managed to identify the vortex ring, but the RBF is shown to be less noisy while allowing for computation of various properties on a much finer grid than AGW without repeating interpolation. It should also be noted that the AGW interpolation shows the vortex ring to be more symmetric, although noisier, while the RBF results shows the vortex to be asymmetric.

The 'FlowFit' algorithm was proposed by Gesemann [9] and used in STB by Schanz *et al* [10]. The algorithm models each component of the flow field using a weighted sum of three-dimensional and evenly spaced quadratic B-splines. In order to evaluate the flow field at arbitrary coordinates, the velocity and acceleration of particles must be known at discrete locations. Systems of equations are defined for each known velocity measurement, and additional equations used to regularize the

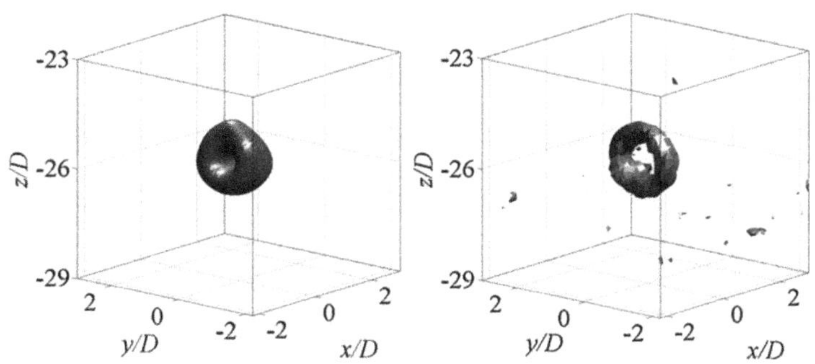

Figure 8.7. Isovorticity surfaces using RBF at $\omega_n = 0.5$ (left) and AGW at $\omega_n = 0.25$ (right) [8].

flow penalize nonzero divergence and nonzero curvatures. A conjugate-gradient algorithm is used to iteratively solve the systems of equations. Then the flow field is sampled on a regular grid, including both velocity and spatial derivative information. Typically 5–20 grid points are generated per tracer particle.

Schneiders *et al* [11, 12] proposed the vortex-in-cell (VIC) and vortex-in-cell-plus (VIC+) methods, which similarly interpolate vectors onto a high-resolution grid. While VIC uses the vorticity-transport equation to interpolate data, the VIC+ method includes the material derivative. The authors suggest generating 64 grid nodes per tracer particle for the VIC+ interpolation. RBFs are used with a window size, $\sigma = 1.1h$, where h is the grid node spacing. The vorticity, $\boldsymbol{\omega}_h$, is defined on the grid points using RBFs and is related to velocity, $\boldsymbol{u}_h$, by

$$\nabla^2 \boldsymbol{u}_h = -\nabla \times \boldsymbol{\omega}_h, \tag{8.11}$$

with boundary conditions on velocity. Second-order central differences are used for calculating spatial derivatives, except at the boundaries, where first-order single-sided differences are used. The vorticity-transport equation is used to calculate the temporal vorticity derivative

$$\frac{\partial \boldsymbol{\omega}_h}{\partial t} = (\boldsymbol{\omega}_h \cdot \nabla)\boldsymbol{u}_h - (\boldsymbol{u}_h \cdot \nabla)\boldsymbol{\omega}_h. \tag{8.12}$$

With this, the temporal velocity derivative is calculated from the temporal vorticity derivative using

$$\nabla^2 \frac{\partial \boldsymbol{u}_h}{\partial t} = -\nabla \times \frac{\partial \boldsymbol{\omega}_h}{\partial t}, \tag{8.13}$$

with boundary conditions on the temporal velocity derivative. This is then integrated to provide $\frac{\partial \boldsymbol{u}_h}{\partial t}$, which is used to calculate the velocity material derivative, defined on the grid as

$$\frac{D\boldsymbol{u}_h}{Dt} = \frac{\partial \boldsymbol{u}_h}{\partial t} + (\boldsymbol{u}_h \cdot \nabla)\boldsymbol{u}_h. \tag{8.14}$$

Optimization is performed to minimize the difference between the original velocity and acceleration measurements and estimates using linear interpolation from the computational grid.

The VIC and VIC+ interpolation algorithms were compared with AGW, linear interpolation (LIN), and PIV results for a simulated turbulent boundary layer. The normalized turbulent statistics for the largest reported particles per pixel ($N_{\text{ppp}} = 0.045$) are given in figure 8.8. PIV has the greatest damping of turbulent fluctuations, while VIC+ most closely matches the DNS results. VIC+ has a substantial improvement even over the standard VIC technique with a computational time that is approximately four times longer than for PIV cross-correlation.

To improve spatial gradient estimations for unstructured Lagrangian datasets, Wong *et al* [13] proposed using the vorticity-transport equation,

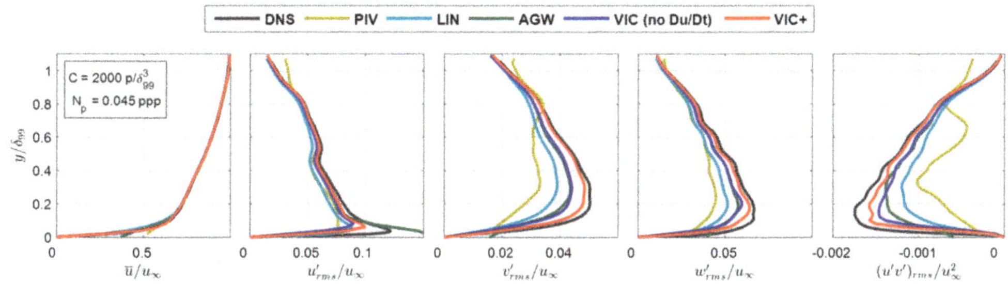

Figure 8.8. Turbulence statistics from different velocity field reconstructions in a turbulent boundary layer. Reprinted from [12] with permission of Springer.

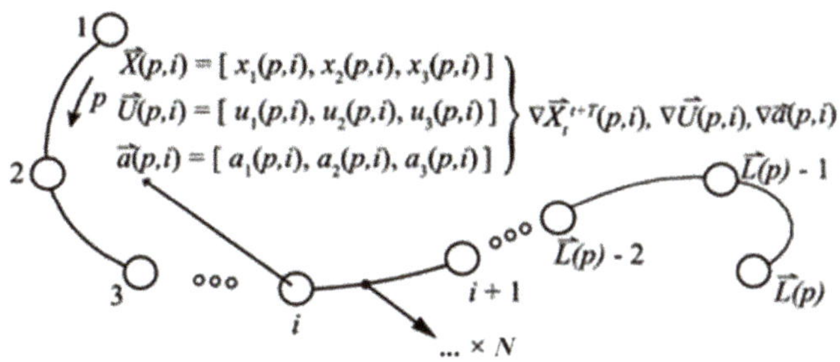

Figure 8.9. A schematic of one of N pathlines acquired within a PTV dataset. The number of time steps within each pathline are contained within the $1 \times N$ vector array $\mathbf{L}$, while the pathlines themselves are enumerated by counter p. As such, the number of time steps in which the current pathline is observed is given by $\mathbf{L}(p)$. The time steps along pathline p are enumerated by the counter i, and at each time step the three-dimensional position $\mathbf{X}(p, i)$, velocity $\mathbf{U}(p, i)$, and acceleration $\mathbf{a}(p, i)$ have been measured. With this dataset, the velocity-gradient tensor $\nabla\mathbf{U}(p, i)$ is to be calculated. Reprinted from [13] with permission of Springer.

$$\frac{D\boldsymbol{\omega}}{Dt} = (\boldsymbol{\omega} \cdot \nabla)\boldsymbol{u} + \nu\nabla^2\boldsymbol{\omega}, \tag{8.15}$$

as a physical constraint, in order to take advantage of its temporal information, i.e. its substantial derivative, along a pathline. Here, they assume that viscous diffusion is negligible since its timescales are greater than those of the measurements in which this approach will be used. Its nomenclature is briefly described in figure 8.9. To implement the constraint, the vorticity-transport equation should not have any residuals at any point nor time. Instead of minimizing the residuals,

$$R(p, i) = \sum_{j=1}^{3}\left(\frac{D\omega_j(p, i)}{Dt} - (\boldsymbol{\omega}(p, i) \cdot \nabla)u_j(p, i)\right)^2, \tag{8.16}$$

individually at each point, the residual across an entire pathline,

$$\sum_{i=1}^{\mathrm{L}(p)} R(p, i) = \sum_{i=1}^{\mathrm{L}(p)}\left[\sum_{j=1}^{3}\left(\frac{D\omega_j(p, i)}{Dt} - (\boldsymbol{\omega}(p, i) \cdot \nabla)u_j(p, i)\right)^2\right], \tag{8.17}$$

is simultaneously minimized in order to maximize the amount of temporal information transfer to the spatial gradient estimation.

To experimentally test this approach, a 30 cm diameter circular disk was linearly accelerated orthogonal to the flow through a $1 \times 1 \times 15\ m^3$ tow tank facility. Particle images were acquired by four Photron SA4 high-speed cameras at 900 Hz. These cameras imaged a $10 \times 10 \times 0.3\ cm^3$ volume, capturing the shear layer shedding from the edge of the disk. Data processed with this method on an instantaneous snapshot clearly shows the rollers within shear layer, as shown in figure 8.10.

8.2 Pressure calculation

Determining a fluid pressure field can help identify flow structures, calculate surface forces, and give insight into sound generation. Velocimetry results can be used to determine the pressure gradient using the momentum equation,

$$\nabla p = -\rho \frac{D\boldsymbol{u}}{Dt} + \mu \nabla^2 \boldsymbol{u}. \tag{8.18}$$

In the case of inviscid flow, the dissipation term can be neglected, giving the pressure gradient as a function of only density and the material derivative. For incompressible flow, the density can be considered constant; if flow is compressible, the equation can be rewritten to eliminate density as an independent variable assuming adiabatic flow.

Interpolation techniques like FlowFit and VIC+ provide material derivative estimates on a fine grid using time-resolved Lagrangian particle tracks. Thus, once boundary conditions have been defined, the pressure field can be determined using path integration. The path integration is generally averaged over many paths to reduce the impact of noise [14].

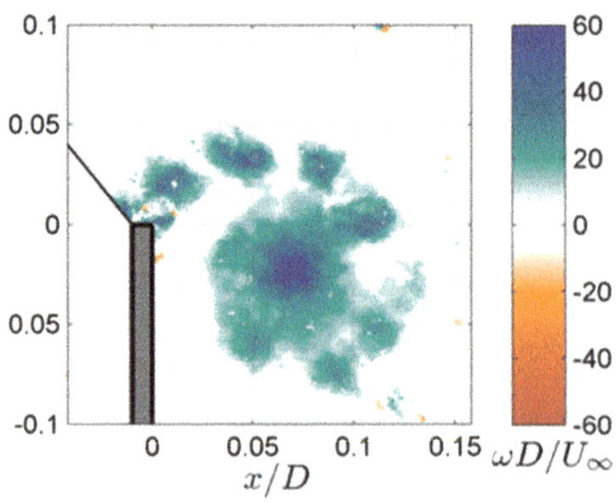

Figure 8.10. An instantaneous snapshot of the vorticity field captured from a single image sequence is shown here after vorticity correction on flow-compiled PTV data. The correction scheme allows for the identification of individual instabilities in the shear layer, and the clear identification of the vortex core. Reprinted from [13] with permission of Springer.

Alternatively, the pressure can be presented in the form of the Poisson equation,

$$\nabla^2 p = \nabla \cdot \left(-\rho \frac{D\boldsymbol{u}}{Dt} + \mu \nabla^2 \boldsymbol{u}\right). \tag{8.19}$$

The Poisson equation can also be modified using the Reynolds-averaged approach given by van Oudheusden [14],

$$\nabla^2 \overline{p} = -\rho \nabla \cdot (\overline{\boldsymbol{u}} \cdot \nabla)\overline{\boldsymbol{u}} - \rho \nabla \cdot \nabla \cdot (\overline{\boldsymbol{u}'\boldsymbol{u}'}). \tag{8.20}$$

This modification allows for ensemble averaging to give higher spatial resolution and turbulent statistics. From this, the Poisson equation can be solved iteratively once boundary conditions have been defined [15].

A technique proposed by Huhn *et al* [16] suggests that the pressure gradient equation can be solved using fast Fourier transform (FFT) integration. The three components of the pressure gradient were introduced as

$$\partial_x P = -\rho a_x, \tag{8.21}$$

$$\partial_y P = -\rho a_y, \tag{8.22}$$

$$\partial_z P = -\rho a_z. \tag{8.23}$$

This is converted to Fourier space such that

$$\widetilde{P}(\boldsymbol{k}) = \frac{\boldsymbol{k} \cdot \widetilde{\nabla P}}{i\,|\boldsymbol{k}|^2} = \frac{k_x \widetilde{\partial_x P} + k_y \widetilde{\partial_y P} + k_z \widetilde{\partial_z P}}{i(k_x^2 + k_y^2 + k_z^2)}, \tag{8.24}$$

and then converted back to normal space

$$P'(\boldsymbol{x}) = FT^{-1}\widetilde{P}(\boldsymbol{k}), \tag{8.25}$$

where $\boldsymbol{k}$ is the wavenumber vector. For $\boldsymbol{k} = 0$, there exists a singularity, so the constant component of the pressure gradient is set to zero. Therefore, the pressure gradient must be reconstructed as

$$P(\boldsymbol{x}) = P'(\boldsymbol{x}) + \langle \partial_x P \rangle x + \langle \partial_y P \rangle y + \langle \partial_z P \rangle z, \tag{8.26}$$

where $\langle \cdot \rangle$ is the spatial mean over the entire domain. The authors suggest enforcing continuous pressure through the boundaries by mirroring pressure across the boundaries. The proposed FFT pressure integration algorithm is completed by adding a constant offset pressure to obtain absolute pressure.

Neeteson and Rival [17, 18] proposed pressure-field extraction directly from unstructured flow data instead of interpolating to a grid. The inviscid form of the Poisson equation was used, and an unstructured mesh was generated using Voronoi tessellation and Delaunay tessellation. For a point i, the Voronoi cell is shown with neighbors j in figure 8.11. Two length measures are used to define the mesh: the distance from point i to point j, h_{ij} and the shared side length of the Voronoi cell, s_{ij}. In 3D space, s_{ij} is the shared cell face area, and additionally a normal vector $\hat{\boldsymbol{n}}_{ij}$ can

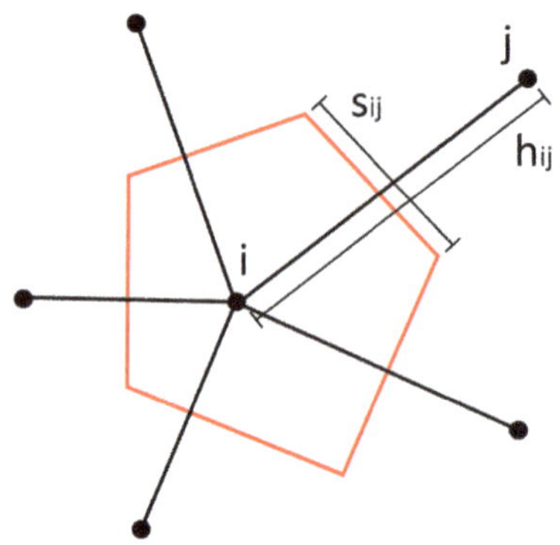

Figure 8.11. Example of Voronoi tessellation in a two-dimensional space [17].

be determined for every connection. Near the boundary, some Voronoi cell faces will be infinitely large, thus these unbounded connections are ignored, and boundary conditions are imposed instead.

Three operators are needed to numerically solve the Poisson equation: the gradient operator, the divergence operator, and the Laplace operator. An expression for evaluating the gradient of a scalar field in an unstructured mesh is

$$\nabla\psi_i = \frac{\sum_j[(\psi_i + \psi_j)s_{ij}\hat{\boldsymbol{n}}_{ij}]}{1/D\sum_j(s_{ij}h_{ij})}, \tag{8.27}$$

and the divergence of a vector is given as

$$\nabla \cdot \boldsymbol{F}_i = \frac{\sum_j[s_{ij}\hat{\boldsymbol{n}}_{ij} \cdot (\boldsymbol{F}_i + \boldsymbol{F}_j)]}{1/D\sum_j(s_{ij}h_{ij})}, \tag{8.28}$$

where D is the dimension of the space. The Laplace operator of a scalar is then given as

$$\nabla^2 f_i = \frac{\sum_j\left(\frac{s_{ij}}{h_{ij}}f_j\right) - f_i\sum_j\left(\frac{s_{ij}}{h_{ij}}\right)}{1/2D\sum_j(s_{ij}h_{ij})}. \tag{8.29}$$

These equations are used to iteratively solve the Poisson equation and enforce Neumann boundary conditions. The iterative integration technique used is known as the successive over-relaxation algorithm, which can be used to stabilize a diverging iterative process or speed up convergence of a slowly converging iterative process.

An experimental comparison of these pressure-field reconstruction techniques was done by van Gent *et al* [19] using a simulated flow over an axisymmetric step flow using zonal detached eddy simulation (ZDES). The model main body has a diameter of 50 mm and after the 15 mm step height, the afterbody has a diameter of

Table 8.1. Global pressure RMS error for PTV pressure extraction techniques for clean and noisy data [19].

Pressure-field reconstruction	Global pressure error (10^{-2})	
	Clean	Noisy
VIC+	0.68	0.63
FFA	0.71	0.74
FFB	0.72	0.76
ILAG	0.88	0.89
VOR	1.71	1.63

20 mm. The freestream Mach number was 0.7 and the total pressure and temperature were 200 kPa and 285 K, respectively. The measurement domain was $60 \times 24 \times 4\ \text{mm}^3$. Both clean and noisy datasets were processed using shake-the-box. The RMS velocity errors in clean and noisy data were 0.01 and 0.04 voxels, respectively.

The PTV algorithms used include FlowFit A (FFA), FlowFit B (FFB), VIC+, iterative pseudo-Lagrangian approach (ILAG) that uses an iterative least-square pseudo-tracking method, and Voronoi-based pressure integration (VOR). The FFA algorithm uses quadratic B-splines for interpolation and performs pressure integration in the Fourier space. FFB generates cubic B-spline functions for the scalar pressure field and estimates its parameters by minimizing a weighted sum of the squared error with measured pressure gradients at particle locations, squared error at the boundary conditions, and squared curvatures of the pressure function. This minimization is a linear least-squares problem that is solved iteratively. The VIC+ algorithm iteratively solves the Poisson equation using the interpolated material derivatives. ILAG uses a first-order least-square fit of velocity to create imaginary particle tracks to determine the material derivative throughout the measurement volume. Spatial integration of the material derivative is then performed to determine the pressure field.

In order to compare the algorithms, a global RMS error was defined as

$$\varepsilon_{\text{r.m.s.}}^{\text{global}} = \sqrt{\frac{1}{N_t N_p} \sum_{i=1}^{N_t} \sum_{j=1}^{N_p} ((C_p)_{i,j} - (C_{p,\,\text{ref}})_{i,j})^2}, \quad (8.30)$$

where N_p is the number of grid points and N_t is the number of time instances. The global errors for the tested techniques are shown in table 8.1. Note that the error is expressed in 0.01 C_p, which is equivalent to a percentage of the freestream dynamic pressure. The global pressure fluctuation level is 4% of the local dynamic pressure, thus a 1% error corresponds to 25% of the global pressure fluctuations. The results show that the global pressure error was lowest using VIC+, with only a 0.63% error for noisy data. The FlowFit techniques performed nearly as well as VIC+, although the Voronoi pressure integration method had an error of 1.63%, more than double that of the VIC+ and FlowFit methods.

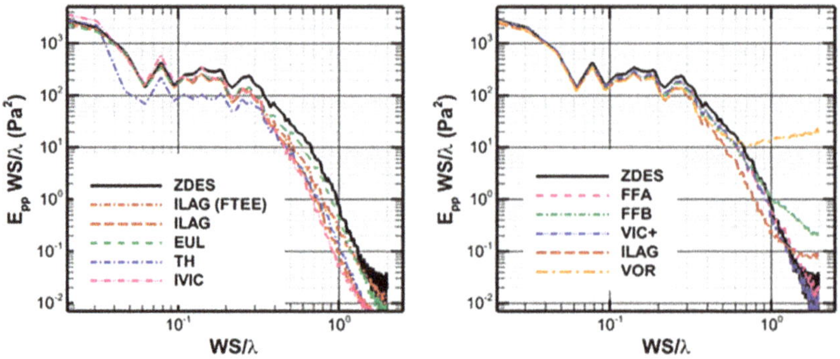

Figure 8.12. Comparison of the spatial spectra of pressure fluctuations along 40 horizontal lines within the shear layer calculated using PIV (left) and PTV (right) algorithms [19].

The spatial spectra of pressure reconstructions were compared by taking the pressure measurements in straight, horizontal lines across the measurement domain and calculating the Fourier transform. Forty horizontal lines were used for each measurement technique and the Fourier coefficients (E_{pp}) were averaged and scaled by the ratio of the window size and the wavelength (WS/λ). Figure 8.12 shows the pressure fluctuation intensities at wavelengths up to the Nyquist frequency ($WS/\lambda = 2$). The left figure shows the pressure-field spectra from PIV measurements, while the right figure shows the pressure-field spectra from PTV measurements. In comparison to the simulated results (ZDES), it can be seen that the PIV approaches are able to generally capture the relative spatial spectral variations, although not matching their amplitudes well. The PTV approaches, on the other hand, are better able to capture the spectral behavior well until the algorithm's failure to resolve the smallest flow features is reached. For example, the VOR, FFB, and ILAG approaches reach this limit at WS/λ = 0.7, 1.0, and 1.25, respectively. The figure also shows that ZDES simulations may have reached its limit at WS/λ = 1.25, where its curve flattens out and its fluctuation begins to increase. The FFA and VIC+ approaches, however, show that they are not only able to capture the smallest scales' spectra quite well, but that they extend beyond the limitations of the ZDES simulations.

References

[1] Adrian R J and Westerweel J 2011 *Particle Image Velocimetry* (New York: Cambridge University Press)

[2] Cohn R K and Koochesfahani M M 2000 The accuracy of remapping irregularly spaced velocity data onto a regular grid and the computation of vorticity *Exp. Fluids* **29** S61–9

[3] Agui J C and Jimenez J 1987 On the performance of particle tracking *J. Fluid Mech.* **185** 447–68

[4] Malik N A and Dracos T 1995 Interpolation schemes for three-dimensional velocity fields from scattered data using Taylor expansions *J. Comput. Phys.* **119** 231–43

[5] Ido T, Murai Y and Yamamoto F 2002 Postprocessing algorithm for particle-tracking velocimetry based on ellipsoidal equations *Exp. Fluids* **32** 326–36

[6] Ido T and Murai Y 2006 A recursive interpolation algorithm for particle tracking velocimetry *Flow Meas. Instrum.* **17** 267–75
[7] Vedula P and Adrian R J 2005 Optimal solenoidal interpolation of turbulent vector fields: application to PTV and super-resolution PIV *Exp. Fluids* **39** 213–21
[8] Casa L D C and Krueger P S 2013 Radial basis function interpolation of unstructured, three-dimensional, volumetric particle tracking velocimetry data *Meas. Sci. Technol.* **24** 065304
[9] Gesemann S 2015 From particle tracks to velocity and acceleration fields using B-splines and penalties, arXiv 1510.09034
[10] Schanz D, Gesemann S and Schroder A 2016 Shake-the-box: Lagrangian particle tracking at high particle image densities *Exp. Fluids* **57** 70
[11] Schneiders J F G, Dwight R P and Scarano F 2014 Time-supersampling of 3D PIV measurements with vortex-in-cell simulation *Exp. Fluids* **55** 1692
[12] Schneiders J F and Scarano F 2016 Dense velocity reconstruction from tomographic PTV with material derivatives *Exp. Fluids* **57** 139
[13] Wong J G, Rosi G A, Rouhi A and Rival D E 2017 Coupling temporal and spatial gradient information in high-density unstructured Lagrangian measurements *Exp. Fluids* **58** 140
[14] van Oudheusden B W 2013 PIV-based pressure measurement *Meas. Sci. Technol.* **24** 032001
[15] Schneiders J F G, Caridi G C A, Sciacchitano A and Scarano F 2016 Large-scale volumetric pressure from tomographic PTV with HFSB tracers *Exp. Fluids* **10** 1007
[16] Huhn F, Schanz D, Gesemann S and Schroder A 2016 FFT integration of instantaneous 3D pressure gradient fields measured by Lagrangian particle tracking in turbulent flows *Exp. Fluids* **57** 151
[17] Neeteson N J and Rival D E 2015 Pressure-field extraction on unstructured flow data using a Voronoi tessellation-based networking algorithm: a proof-of-principle study *Exp. Fluids* **56** 44
[18] Neeteson N J, Bhattacharya S, Rival D E, Michaelis D, Schanz D and Schroder A 2016 Pressure-field extraction from Lagrangian flow measurements: first experiences with 4D-PTV data *Exp. Fluids* **57** 102
[19] Van Gent P L *et al* 2017 Comparative assessment of pressure field reconstructions from particle image velocimetry measurements and Lagrangian particle tracking *Exp. Fluids* **58** 33

Dana Dabiri and Charles Pecora

Chapter 9

Conclusions

Particle tracking velocimetry is becoming a powerful quantitative flow visualization technique, capable of outputting high-resolution vector fields that detail flow features such as those shown in figure 6.5. Recent methodologies in triangulation, tomo-PTV, and shake-the-box methods have been able to process images with 0.035–0.05 N_{ppp} with localization uncertainties as low as 0.1 px and 0.15 px for in-plane and out-of-plane errors, respectively [1], and most recently as high 0.125 N_{ppp} with localization uncertainties of 0.3 px and 0.3 px for in-plane and out-of-plane errors, respectively [2]. Advances in laser and camera technologies now allow for volume illumination and image acquisitions up to 20 kHz, thereby allowing for time-resolved particle velocimetry techniques. Such time-resolved methods are therefore able to provide flow accelerations, and consequently pressure distributions within flow, which are exciting capabilities, although at present, the uncertainties of these approaches tend to be rather high [3]. In addition, with the consequential increased data rate collection, a typical bottleneck for current PTV systems is therefore the increased computational time required to process acquired data.

The sources of error in PTV experiments come from each of the steps involved. First, error inherently exists in the difference between the tracer particle motion and the fluid in which it is suspended. Additionally, in 3D-PTV, the errors associated with proper particle identification and spatial localization, which can be amplified by noise, compile. These errors then propagate throughout velocity calculations, but careful calibration and implementation of the most advanced particle locating algorithms can minimize the reconstruction errors. Finally, particle matching errors can cause loss of data and incorrect velocity vectors. The shake-the-box approach is to resolve these by combining the individual steps for 3D reconstruction and tracking to ensure particle tracks over at least four sequential time steps.

doi:10.1088/978-0-7503-2203-4ch9

Three different open-source PTV software packages using some of the methods discussed in this review have been developed with documentation and instructions for their use. PTVlab[1] is a MATLAB software package for analyzing 2D PTV images. The software uses dynamic threshold binarization [4] and Gaussian centroid estimation [5] to identify and locate particle centers, then tracks particles using a hybrid cross-correlation and relaxation method [6]. The software package also includes vector validation, interpolation on to a regular grid, and streamline calculation. Researchers at ETH Zurich, the TU/e Turbulence and Vortex Dynamics group, and the Turbulence Structure Laboratory at Tel Aviv University contributed to the development of OpenPTV[2], an open-source software for 3D-PTV experiments. OpenPTV software uses the triangulation and tracking algorithms used by Maas *et al* [7], Malik *et al* [8], and Kreizer and Liberzon [9]. Finally, Professor Nicholas Ouellette, at the Civil and Enviornmental Engineering Department at Stanford University, has also made his particle tracking algorithms available[3] [10]. His algorithms use the modified anisotropic thresholding operator developed to identify particle blobs, and two one-dimensional Gaussian fits to each particle blob to identify its position. In addition, he has also used neural networking algorithms to identify particles. Finally, he uses a multi-frame approach to track particles. All resources have instructions that describe the steps for setting up a PTV experiment and implementing the software.

While PTV has flourished significantly, more efficient and accurate systems with increased temporal and spatial resolution can be required for analysis of complex flow structures. These limitations can be removed with CCD sensors capable of higher frame rates and higher resolution, as well as illumination sources that have larger illumination intensities in addition to the higher repetition rates to allow for larger volume imaging. Furthermore, improving localization and tracking algorithms can allow for higher density images to be analyzed with increased accuracy as well as reducing ghost particles. Within the context of time-resolved PTV, this could also lead to more accurate pressure measurements as well. With these improvements, 3D-PTV will lead current researchers into a future where high-resolution and accurate 3D experimental data can be obtained that would better enable us to pursue scientific understanding of three-dimensional fluid flows.

References

[1] Fuchs T, Hain R and Kähler C J 2016 Double-frame 3D-PTV using a tomographic predictor *Exp. Fluids* **57** 174

[2] Schanz D, Gesemann S and Schroder A 2016 Shake-the-box: Lagrangian particle tracking at high particle image densities *Exp. Fluids* **57** 70

[3] Van Gent P L *et al* 2017 Comparative assessment of pressure field reconstructions from particle image velocimetry measurements and Lagrangian particle tracking *Exp. Fluids* **58** 33

[1] PTVlab software and documentation are available at http://ptvlab.blogspot.com.

[2] OpenPTV software and documentation are available at http://www.openptv.net.

[3] https://web.stanford.edu/~nto/LPT.shtml.

[4] Ohmi K and Li H-Y 2000 Particle-tracking velocimetry with new algorithms *Meas. Sci. Technol.* **11** 603–16

[5] Marxen M, Sullivan P E, Loewen M R and Jahne B 2000 Comparison of Gaussian particle center estimators and the achievable measurement density for particle tracking velocimetry *Exp. Fluids* **29** 145–53

[6] Brevis W, Niño Y and Jirka G H 2011 Integrating cross-correlation and relaxation for particle tracking velocimetry *Exp. Fluids* **50** 135–47

[7] Maas H G, Gruen A and Papantoniou D 1993 Particle tracking velocimetry in three-dimensional flows. Part 1. Photogrammetric determination of particle coordinates *Exp. Fluids* **15** 133–46

[8] Malik N A, Dracos T and Papantoniou D A 1993 Particle tracking velocimetry in three-dimensional flows. Part 2. Particle tracking *Exp. Fluids* **15** 279–94

[9] Kreizer M and Liberzon A 2011 Three-dimensional particle tracking method using FPGA-based real-time image processing and four-view image splitter *Exp. Fluids* **50** 613–20

[10] Ouellette N, Xu H and Bodenschatz E 2006 A quantitative study of three-dimensional Lagrangian particle tracking algorithms *Exp. Fluids* **40** 301–13

www.ingramcontent.com/pod-product-compliance
Ingram Content Group UK Ltd.
Pitfield, Milton Keynes, MK11 3LW, UK
UKHW051854100926
8556IPUK00005B/617